CLIMATE CHANGE AND ITS CAUSES,
EFFECTS AND PREDICTION

CLIMATIC EFFECTS CREATED BY ATMOSPHERIC GREENHOUSE GASES

CLIMATE CHANGE AND ITS CAUSES, EFFECTS AND PREDICTION

Additional books in this series can be found on Nova's website under the Series tab.

ENVIRONMENTAL SCIENCE, ENGINEERING AND TECHNOLOGY

Additional books in this series can be found on Nova's website under the Series tab.

CLIMATE CHANGE AND ITS CAUSES,
EFFECTS AND PREDICTION

CLIMATIC EFFECTS CREATED BY ATMOSPHERIC GREENHOUSE GASES

ALEXANDER Y. GALASHEV

Nova Science Publishers, Inc.
New York

Copyright © 2010 by Nova Science Publishers, Inc.

All rights reserved. No part of this book may be reproduced, stored in a retrieval system or transmitted in any form or by any means: electronic, electrostatic, magnetic, tape, mechanical photocopying, recording or otherwise without the written permission of the Publisher.

For permission to use material from this book please contact us:
Telephone 631-231-7269; Fax 631-231-8175
Web Site: http://www.novapublishers.com

NOTICE TO THE READER

The Publisher has taken reasonable care in the preparation of this book, but makes no expressed or implied warranty of any kind and assumes no responsibility for any errors or omissions. No liability is assumed for incidental or consequential damages in connection with or arising out of information contained in this book. The Publisher shall not be liable for any special, consequential, or exemplary damages resulting, in whole or in part, from the readers' use of, or reliance upon, this material. Any parts of this book based on government reports are so indicated and copyright is claimed for those parts to the extent applicable to compilations of such works.

Independent verification should be sought for any data, advice or recommendations contained in this book. In addition, no responsibility is assumed by the publisher for any injury and/or damage to persons or property arising from any methods, products, instructions, ideas or otherwise contained in this publication.

This publication is designed to provide accurate and authoritative information with regard to the subject matter covered herein. It is sold with the clear understanding that the Publisher is not engaged in rendering legal or any other professional services. If legal or any other expert assistance is required, the services of a competent person should be sought. FROM A DECLARATION OF PARTICIPANTS JOINTLY ADOPTED BY A COMMITTEE OF THE AMERICAN BAR ASSOCIATION AND A COMMITTEE OF PUBLISHERS.

Additional color graphics may be available in the e-book version of this book.

LIBRARY OF CONGRESS CATALOGING-IN-PUBLICATION DATA
Galashev, Alexander Y.
  Climatic effects created by atmospheric greenhouse gases / Alexander Y. Galashev.
    p. cm.
  Includes index.
  ISBN 978-1-61761-661-7 (softcover)
  1. Greenhouse gases. 2. Greenhouse effect, Atmospheric. 3. Climatic changes. 4. Human beings--Effect of climate on. I. Title.
  TD885.5.G73G35 2011
  551.51'12--dc22
                                                        2010031175
  ISBN: 978-1-61761-661-7

Published by Nova Science Publishers, Inc. † New York

CONTENTS

Preface		vii
Abbreviations		xi
Chapter 1	Introduction	1
Chapter 2	The Greenhouse Gas Clusterization	7
Chapter 3	Cluster Mechanism of Ozone Destruction by Chlorine and Bromine Ions	23
Chapter 4	Estimation of the Greenhouse and Anti-Greenhouse Effects	41
Chapter 5	Concluding Remarks	63
References		67
Index		73

Preface

The temperature on the Earth has exceeded the trend by 0.6 degrees for the last three decades and by 0.74 degrees for the last century. Since 1975 the accelerated warming - of ~ 0.2 K per one decade has been observed. According to the forecast of the Main astronomical observatory of the Russian Academy of Sciences, in the nearest years we should expect a global downturn of temperature due to a reduction of full stream sunlight. In 50 years a condition of deep cold snap will occur, and new warming will begin only in the beginning of XXII century. It is connected with approach of the next 200-years global warming cycle. The opposite forecast was given by the American Geophysical Union (AGU). According to the AGU during last years the ice cover of Arctic regions experiences fast and, probably, irreversible changes. By 2040 the Arctic Ocean finally will lose an all-year-round ice cover, and by 2080 population of Europe will no longer experience snowy winters. Thus, forecasts concerning the change of climate are relatively inconsistent. Measures on stabilization of the climate appear extremely expansive. In this situation it is difficult to hope for support of specific measures offered by politicians to decrease the rate of climatic change. However, it is too serious a question to disregard it since the destiny of mankind is remains uncertain. According to the report of the Intergovernmental Panel on Climate Change (IPCC) they forecast rapid thawing of glaciers, gradual disappearance of permafrost and the rise of the world's oceans as a consequence of mid-annual temperature increase. To this end the potential of significant disruption due to climatic change remains a potent issue for the 21st century. According to today's forecasts, to the end of current century the Earth's temperature will increase by minimum 3K. While an increase of 2 K already represents danger. Beyond this temperature climatic change becomes irreversible. How can one convince politicians to rise on the struggle against "the human factor" global warming? First of all in investigating this problem it is necessary to take into account many additional factors including inconsistent conclusions relating to measures of climate change restraint.

Until present, climate change issues of the Earth have not received due attention. This is primarily due to weak evidence relating to the reasons behind modern global warming. Hence, recommendations for restraint within the current climate change framework are ambiguous and are not perceived as urgent. There is insufficient scope of knowledge pertaining to natural phenomena influencing climate change and their interdependence. In the present brochure one of the natural phenomena that previously has not been included in the scope of problems related to the greenhouse effect is considered in detail. In this work attention is drawn to the clusterization of greenhouse gases occurring in the troposphere due to the rapid decrease of temperature commensurate with altitude increase. An initial decrease of temperature comes to an end at an altitude of ~12–15 kms. The basic greenhouse gas is water vapor which creates up to ~ 70 % of all greenhouse effect in the atmosphere. Such role water vapor receives both owing to its higher abundance comparing to carbon dioxide (CO_2) and due to ability of water molecules to absorb infra-red (IR) radiation in wide range. CO_2 gas makes considerably smaller effect than water vapor (but greater than the effect of other greenhouse gases). However its emission as a result of human activity can be easily defined. Water is a unique substance existing in the atmosphere in three aggregate states (gaseous, liquid and solid), and it also forms clusters which can be considered as a special state of substance. Clusters represent fragments from water molecules generated due to hydrogen linking. In the non-dense environment, which the atmosphere mainly is, clusters appear in steady formations. Their time of life appreciably exceeds the characteristic time of absorption and the emission of IR radiation by these objects. In surroundings of supersaturated water vapor (in clouds) clusters grow to the size of drops or crystals, forming deposits. Due to extended surface water clusters have a high ability to absorb. In the atmosphere they can grasp molecules of other gases, including greenhouse ones. As well as other formations from water molecules (vapor of monomers, drops, crystals) atmospheric water clusters create greenhouse effect which by our estimation makes ~ 1.1 ± 0.1 K. Remarkably, water clusters are able to create cooling or an anti-greenhouse effect. Similar properties can be demonstrated by aerosols that in some cases are formed from clusters. The essence of the cluster anti-greenhouse effect is that a small cluster (containing up to 20 water molecules) can absorb energy of IR radiation not bigger than one absorbed by a water molecule. Thus the absorbing ability of α cluster consisting from n water molecules can be reduced by factor n. On average the magnitude α of atmospheric water clusters decreases by 2.4 times, and the anti-greenhouse effect created by them makes ~ 3.3 ± 0.3 K. This value exceeds the increase of global temperature fixed for last 100 years by 4.5 times. The stratospheric ozone absorbs ultra-violet radiation which is pernicious for living organisms. Tropospheric ozone is not only a greenhouse gas, but also the strongest oxidant on basis of which smog is formed. Water clusters play a catalyst role in terms of destruction of ozone by ions of chlorine and bromine on their surface. In the present work it is shown that ions of bromine stay on the surface of water clusters for much longer period, thus creating more favorable conditions for destruction of ozone by ions of bromine. For consideration of climatic changes it is necessary to take into account not only change of content of greenhouse gases in the troposphere but also the quantity of ozone in the stratosphere, and even the atomic oxygen in thermosphere.

Alexander Y. Galashev
Yekaterinburg
July 2010

ABBREVIATIONS

AGU	American Geophysical Union
GHG	Greenhouse Gases
IPCC	Intergovernmental Panel on Climate Change
IR	Infrared
MD	Molecular Dynamics
PBL	Planetary Boundary Layer

Chapter 1

INTRODUCTION

The Sun's activities and the earth's position in relation to the Sun have a direct impact on the climate. In addition to the general heat coming from the sun, there are also sunspots generated on the Sun's surface. These sunspots are areas with increased magnetic activity and seem to be the cause of Sun flares and mass ejections. When Earth is in the path of these flares, enormous waves of additional heat and magnetism influence our planet. Higher temperatures can be result of it, particularly when certain protective layers of the atmosphere disappear. The Earth's orbit around the Sun is elliptical, and so there are times when the Earth is closer to the Sun than others. When it occurs, it is normal to have an increase in global mean surface temperature. The Earth is kept warm by gases in the atmosphere. These gases, such as water vapor, carbon dioxide, methane, nitrous oxide, and ozone (in decreasing order of effectiveness, mainly due to their concentrations) prevent the heat from escaping into space. If our atmosphere contained none of these greenhouse gases (GHG), it would be a very cold place.

Global temperature stably rises. Even a small rise has vast impact on the *environment*. With more heat in the atmosphere the temperature of oceans goes up. This in turn impacts ice caps, as well as hot water eco-systems. At ice caps glaciers begin to melt. Ice caps and oceans play a very important role in balancing the global climate because their condition influences the reflection of sunlight. Inland glaciers also feel the effect and melt. As ice caps and inland glaciers melt, water levels rise. This puts low-lying coastal areas under threat of flooding.

The question of the nature of contemporary global warming cannot be regarded as answered to this day. There is ample reason to believe that the fluctuation pattern of climatic variations is the natural property of climatic development. Therefore, one can use the graph of climatic variations in the previous years, for example, in the last 10 000 years, and predict the climatic development for the distant future. The astronomic concept suggested by M. Milankovich [1] is based on the chronological correspondence between cooling-warming periods and cyclic variation of insolation. This theory explains, with relative accuracy, the long- and medium-period cycles of climatic variation. The emergence of short-period cycles may be associated with the effect of the Earth's magnetic field which causes a variation of the flux of cosmic rays proceeding into and from the atmosphere. No direct evidence is available of the fact that the increase in the global air temperature is caused by the accumulation of GHG in the atmosphere. This may be a manifestation of natural fluctuation. For example, the global temperature somewhat decreased from the 1970s to 1990s in spite of the highest ever level of CO_2. If we assume that the average temperature of the Earth surface is largely defined by convective motion of air masses in the troposphere, the replacement of the present nitrogen-oxygen atmosphere by carbon-dioxide one will in fact cause little change as regards the heat balance. In this case, it may be expected that CO_2 will make a directly opposite effect. In other words, the higher the CO_2 content in the atmosphere, the higher the absorption of the Earth thermal radiation will be. This will result in a more intensive convective exchange of air masses in the troposphere. Therefore, the carryover of abnormal excess of heat to the stratosphere will be still faster. It is quite probable that the ocean is the original cause of the variation of CO_2 content in the atmosphere. This follows from the fact of permanent equilibrium observed between the CO_2 content in the atmosphere and its content in the hydrosphere. Sixty times more CO_2 is dissolved in the ocean than it is contained in the atmosphere. The lower the temperature of water, the more CO_2 is dissolved in water. Hence follows that an increase in CO_2 content will correspond to climatic warming and, vice versa, a decrease in its content – to cooling. Water vapor concentrated in the troposphere, especially in its lower part, is the main absorber of radiation in the atmosphere. Of the overall composition of solar radiation, water vapor absorbs a significant fraction in the infrared (IR) spectral region.

From the beginning of the Industrial Revolution human impact on nature has dramatically increased. The burning of fossil fuels in transport, industry, and power plants produce carbon dioxide and other heat-trapping gasses in quantities comparable with natural. Since more and more forests are being cut down there are less trees to absorb carbon dioxide, resulting in an increase of greenhouse gas content in the atmosphere.

Within a period of global warming, there could be shorter periods of cooling, off-setting the warming to some extent. Obviously we cannot change the natural causes, such as the Earth's orbit around the Sun and the Sun flares. But we can have an impact on the man-made causes. Just as we may have caused an aggravation of warming, we can also take an active role in reducing continued aggravation.

The Earth's atmosphere is a complex dynamic system, which protects the biosphere. One of the significant factors impacting the Earth's radiation balance is the greenhouse effect. Water vapor and atmospheric gases, such as CO_2, CH_4, N_2O, and others, have a decisive influence on the formation of thermal radiation fields. However, according to the Le Chatelier principle, there are opposite compensating processes in the atmosphere. Clusterization of greenhouse gases can be considered as one of these processes. The temperature contributions of the greenhouse gases of the Earth's atmosphere to the greenhouse effect are determined in [2] according to their volume fraction. So for water vapor this contribution should make 37.4 K. However, correction of 13.4 K is necessary to take into account the effect of water evaporation. Therefore effectively the contribution of water vapor decreases to 24.0 K.

The global greenhouse effect may have more to do with atmospheric water than gases such as carbon dioxide and methane [3]. Water vapor causes 36-70% of the greenhouse effect on Earth, not including clouds, while CO_2 causes only 9–26%, and CH_4, only 4–9% [4]. Although increases of CO_2 are indeed a source of the enhanced greenhouse effect, and thus global warming, the contribution of atmospheric water is rarely discussed because, unlike most other gases, the distribution of atmospheric water varies strongly with altitude, terrestrial location, and time, while water vapor changes to the liquid and solid phases at terrestrial temperatures. Water vapor is conventionally viewed as a gas of individual H_2O molecules. However, from spectroscopic studies [5], atmospheric water vapor is known to be a natural source of clusters of water molecules.

Most of the ozone on earth isn't on Earth at all, but in the layer of the Earth's atmosphere called the stratosphere. This is the upper layer of the atmosphere and starts between 12.9 to 19.3 km above our heads and goes upwards to almost 50 km. Ozone is very minor contributor to the greenhouse effect in the troposphere, but has a very important contribution to the properties of the stratosphere. The radiation, known as the electromagnetic spectrum, consist of gamma rays, X-rays, ultra-violet radiation, visible light and infrared radiation, in decreasing order of energy. The vast majority of gamma and X-rays (highly energetic end of radiation from the Sun) are absorbed by upper atmosphere. Ozone in the stratosphere has a positive benefit to life on Earth as it absorbs harmful ultra-violet light from the Sun while letting through other light wavelengths. The ozone hole occurs during the antarctic spring, from September to early December, as strong westerly wind start to circulate around the continent and create an atmospheric container. In this container over 50% of the lower stratospheric ozone is destroyed. Spring brings an increase of ultraviolet light to the lower antarctic stratosphere, providing the energy needed for the rapid catalytic break-down of ozone by ClO and its dimer $ClOOCl$. Another mechanism involving bromine adds extra 33% to the depletion total. As a rule the stratospheric ozone is destroyed by these two mechanisms, most of the damage occures in the lower stratosphere. The antarctic hole is a warning that if conditions become more antarctic: cooler stratospheric temperatures, more stratospheric clouds, more active chlorine, then global ozone will decrease at much greater pace. Some of the most popular senarios of global warming predict that these changes could occur in larger portions of the stratosphere. The radiative cooling caused by the lower stratospheric ozone loss may offset the radiative warming of the ozone-depleting chemicals. The potential radiative impact of the ozone change on the stratosphere–troposphere exchange was investigated in [6]. This analysis shows that a 15% decrease in global O_3 can cause a maximum cooling of 2.4 K in the stratosphere. It was found a steady decline of free oxygen in the Earth atmosphere. Such situation has been expected, since every molecule of additional carbon dioxide locks up two oxygen atoms. However the free oxygen decline is greater than the carbon dioxide lock-up. The greater than expected overall free oxygen decline is proof that the Earth's photosynthetic capacity has declined.

It is well known that rain water contains oxygen and minerals and crystallizes into snow flakes. Natural clearing the atmosphere from gaseous pollution and aerosols occurs due to the water cycle. As a rule, water in the atmosphere is represented in three aggregation states: gas, liquid and solid. Recently, increased attention is focused on clusters which are formed due to hydrogen bonding. Clusters, especially aggregates of small size, can be considered as a special state of substance. Water vapor as the main representative of water in the atmosphere contributes significantly to its clearing. Clusters are formed from water vapor and subsequently they form water drops or snowflakes which fall as precipitation. At their forming stage and during their subsequent presence in the atmosphere, both clusters and larger formations absorb molecules of pollution substances. Because of a high number of molecules on the surface compared to the number of bulk molecules, clusters are the most active absorbents of molecules. Cluster's reaction to external electromagnetic radiation is defined by the frequency of vibrations of the total dipole moment of the cluster. For clusters of smaller sizes (up to 100 molecules) this frequency hardly changes with the change of cluster size. Hence, all clusters have energetically close responses, and the bigger the cluster the lower activity per molecule to radiation it has. As a result, at cluster formation from molecules the ability to absorb and disperse radiation reduces. Hence, as a whole, the clusterization should be accompanied by a cooling effect. Aerosols also contribute to clearing of the atmosphere: they absorb gaseous pollution, and then together with the deposits which they have absorbed, come back to the Earth. The GHG change by itself produces warming that exceeds that observed by some 40 % on average [7]. Tropospheric aerosols reduce the influence from greenhouse gases. Cooling associated with aerosols restore greenhouse effect to the observed level. This estimation however has large uncertainties in aerosol amount, composition, and physical and optical properties used in modeling of atmospheric aerosols.

Influence of water clusters on formation of IR spectra and their participation in creation of a greenhouse effect has been discussed for a long time [8-11]. Population of the atmosphere by water dimers is supposed at modeling of water recondensation [12]. Calculations specify, that water dimers can exist in significant concentration ($\sim 10^{16}$ cm^{-3} at 313 K and 100 % relative humidity) and influence physical and chemical processes in the atmosphere [13]. The spectra of air received with the use of ionic spectrometer, show the presence of ionic clusters of the size about 1 nm with almost constant concentration [14]. Until recently the existence of atmospheric clusters not carrying electric charge (neutral) was represented as an open question. It is caused by the difficulty of their direct detection. Experimental confirmation of the presence of neutral clusters in the atmosphere demands usage of unique counters for condensation particles [15]. The usage of equipment for cross chemical ionization in experiments on nucleation in ternal system already has revealed a big number of neutral clusters [16]. Water clusters that absorbed

SO_3 molecules prove themselves by catalytic effect at formation of sulfuric acid [17] and by formation of liquid aerosols in the atmosphere. A quantum chemistry study of smaller water clusters, such as trimers, tetramers, and pentamers, has concluded that they too may contribute significantly to global warming [18].

Water vapour is itself too strong a GHG. However the greenhouse effect will decrease at presence of the clusters formation in water vapour instead of to amplify. The increase of the water vapour content in atmosphere at the present condition of an atmosphere results in amplification of a greenhouse effect, but not in direct ratio to quantity added vapour. Now clusters are formed and they reduce a greenhouse effect. However there can come the moment of such warming in atmosphere, that clusters will not be formed. It will result to significant jump of global temperature and its further very fast growth, i.e. to catastrophe. The autoregulation of the atmosphere composition is due to the formation of water clusters and their subsequent capture of greenhouse gas molecules is, as a rule, ignored in the estimation of the Earth's radiation balance.

The goal of our work is to investigate the interaction of water clusters with important GHG, to show influence of this interaction on spectral characteristics of disperse water medium, to establish the role of water clusters in thermal balance of the atmosphere, and also to define their influence on depletion of ozone layer.

Chapter 2

THE GREENHOUSE GAS CLUSTERIZATION

2.1. COMPUTER MODEL

Water is among the most studied of chemicals, owing, in part, to its ubiquity and its necessity for all life. In addition to these "natural" reasons, water is an interesting compound because it has unique physical properties and is a model hydrogen-bonded liquid. Most of the available water interaction potentials are parametrized to reproduce the thermodynamic and structural properties of bulk water. The polarizable model allowed us to examine the changes in the dipole moment of the individual water molecules as a function of their environment. This can provide insight into many-body effects in water clusters at a molecular level. Dang and Chang [19] have developed a polarizable potential model for water that behaves reasonably well with changes in the environments (i.e.., cluster, liquid, and liquid/vapor). The total interaction energy of the system can be written as

$$U_{tot} = U_{pair} + U_{pol},$$

where the pairwise additive part of the potential is the sum of the Lennard-Jones and Coulomb interactions,

$$U_{pair} = \sum_i \sum_j \left[4\varepsilon^{(LJ)} \left\{ \left(\frac{\sigma_{ij}^{(LJ)}}{r_{ij}} \right)^{12} - \left(\frac{\sigma_{ij}^{(LJ)}}{r_{ij}} \right)^{6} \right\} + \frac{q_i q_j}{r_{ij}} \right].$$

Here, r_{ij} is the distance between site i and j, q is the charge, $\sigma^{(LJ)}$ and $\varepsilon^{(LJ)}$ are the Lennard-Jones parameters.

The nonadditive polarization energy is given by

$$U_{pol} = -\frac{1}{2}\sum_i \mathbf{d}_i \cdot \mathbf{E}_i^0,$$

where $\mathbf{E}_i^0$ is the electric field at site i produced by the fixed charges in the system

$$\mathbf{E}_i^0 = \sum_{j \neq i} \frac{q_j \mathbf{r}_{ij}}{r_{ij}^3},$$

d_i is the induced dipole moment at atom site i, and is defined as

$$\mathbf{d}_i = \alpha_i^p \cdot \mathbf{E}_i,$$

where

$$\mathbf{E}_i = \mathbf{E}_i^0 + \sum_{j \neq i} \mathbf{T}_{ij} \cdot d_j.$$

In the above, $\mathbf{E}_i$ is the total electric field at atom i, α_i^p is the atomic polarizability, and $\mathbf{T}_{ij}$ is the dipole tensor

$$\mathbf{T}_{ij} = \frac{1}{|r_{ij}|^3}(3\hat{\mathbf{r}}_{ij}\hat{\mathbf{r}}_{ij} - \mathbf{1}). \tag{1}$$

In (1), $\hat{\mathbf{r}}_{ij}$ is the unit vector in the direction $\mathbf{r}_i - \mathbf{r}_j$, where $\mathbf{r}_i$ and $\mathbf{r}_j$ are the positions of the centers of mass of molecules i and j, and $\mathbf{1}$ is the 3 × 3 unit tensor.

The simulation of water clusters was performed using a refined TIP4P interaction potential for water and the rigid four-center model of H_2O molecule [20]. The modification of interaction potential for water by Dang and Chang [19] concerned the variation of the parameters of the Lennard–Jones part of potential and localization of negative charge. As a result, the value of permanent dipole moment for water molecule was taken to be equal to its experimentally obtained value of 1.848 D. The geometry of this molecule corresponds to the experimentally obtained parameters of the molecule in the gas phase: r_{OH} = 0.09572 nm and angle HOH of 104.5° [21]. Fixed charges (q_H = 0.519 e, q_M = −1.038 e) are ascribed to H atoms and to point M lying on the bisectrix of angle HOH at a distance of 0.0215 nm from oxygen atom. The values of charges and the position of point M are selected so as to reproduce the experimentally obtained values of dipole and quadrupole moments [22, 23], as well as the ab initio calculated energy of dimer and the typical distances in the dimer [24]. The stabilization of short-range order in water clusters is largely attained owing to the short-range Lennard-Jones potential with the center of interaction ascribed to the oxygen atom. Related to point M in addition to the electric charge is the polarizability which is required for the description of nonadditive polarization energy. The standard iterative procedure is used at every time step for calculating induced dipole moments [19]. The accuracy of determination of $\mathbf{d}_i$ is given in the range of $10^{-5}-10^{-4}$ D.

The atom-atom impurity (carbon, oxygen, hydrogen, nitrogen)–water interactions were preassigned in terms of the sum of repulsion, dispersion, and Coulomb contributions,

$$\Phi(r_{ij}) = b_i b_j \exp[-(c_i + c_j)r_{ij}] - a_i a_j r_{ij}^{-6} + \frac{q_i q_j}{r_{ij}}, \qquad (2)$$

where the parameters a_i, b_i, and c_i of the potential describing these interactions were borrowed from Spackman [25, 26]. Experimental polarizability values have been used [27].

Based on the results of high-level ab initio calculations, it was demonstrated [20] that the most energetically favorable structure of $(H_2O)_{20}$ cluster is formed on the basis of pentagonal prism (U_{tot} = -(9.32–9.44) eV) rather than of dodecahedron (-8.67 eV) or fused cubes (-9.21 eV). Molecular dynamics (MD) calculations performed using various empirical potentials are still incapable of producing an unambiguous answer to the question of which structure of $(H_2O)_{20}$ cluster corresponds to the lowest energy. The internal energy of $(H_2O)_{20}$ cluster in the model given by us is -8.66 eV.

The Gear method of the fourth order [28] was used for determining the trajectories of the centers of mass of molecules. The integration time step Δt was 10^{-17} s. First, in a MD calculation with a duration of $2 \times 10^6 \Delta t$, the equilibrium state at $T = 233$ K was prepared for water clusters which contained no impurity molecules. The maximum of isothermal compressibility of supercooled water is observed at this temperature [29] and, therefore, it exhibits the lowest mechanical stability. The clusters which inherit the properties of liquid water must likewise be least stable at $T = 233$ K. Therefore, successful simulation at this temperature presumes the possibility of obtaining stable clusters at other temperatures as well ($T < 273$ K).

The $(H_2O)_{20}$ cluster configuration relating to the time instant of 20 ps was subsequently used as initial configuration for simulating $X_i(H_2O)_n$ clusters where $X = CO_2$, CH_4 or N_2O. Each impurity molecule added was firstly arranged so that the minimal distance between atoms of this molecule and atoms of water molecules would be about 0.6 nm. The molecule of CO_2 or N_2O was arranged so that its axis coincided with the line connecting the center of mass of $(H_2O)_n$ cluster to the center of mass of this (admixture) molecule. Initial orientation of the CH_4 molecule was random. The cut-off radius r_c of all interactions in the model was 0.9 nm. The newly formed cluster was balanced during the time interval of $0.6 \times 10^6 \Delta t$ at $T = 233$ K, and then other necessary physicochemical properties were calculated at the same temperature during the interval of $2.5 \times 10^6 \Delta t$. The Rodriguez-Hamilton parameters [30] were used to derive the analytical solution of equations of motion for molecular rotation, and the scheme of integration of the equation of motion in the presence of rotations corresponded to the approach suggested by Sonnenschein [31].

2.2. DIELECTRIC PROPERTIES

At first let us consider absorption of GHG molecules by small water clusters. Larger clusters will be considered in Chapter 3 in connection with the research of cluster mechanism of ozone destruction. Four systems of clusters were investigated, namely, I – system consisting of $(H_2O)_n$ clusters of the size from 2 up to 20 molecules, , II – $(CO_2)_i(H_2O)_{10}$ clusters, III – $(CH_4)_i(H_2O)_{10}$ clusters, and IV – $(N_2O)_i(H_2O)_{10}$ clusters, $i = 1, ..., 10$. It was assumed that a cluster containing i molecules of CO_2, CH_4 or N_2O and n molecules of water has the statistical weight of

$$W_{i,n} = \frac{N_{i,n}}{N_{i,\Sigma}}, \qquad i = 1, ..., n,$$

where $N_{i,n}$ is the number of clusters with i molecules of impurity and n molecules of water per 1 cm^3, $N_{i,\Sigma} = \sum_{i=1}^{n} N_{i,n}$. The value of $N_{i,n}$ was estimated as follows. We will consider the case of scattering of nonpolarized light, where the molecular free path l is much less than the wavelength of light λ. The extinction (attenuation) coefficient h of incident beam may be determined, on the one hand, by the Rayleigh formula [32] and, on the other hand, in terms of the scattering coefficient η [33] in the approximation of scattering at an angle of 90°. In view of the fact that $h = \alpha + \eta$, where α is the absorptance, we have

$$N_{i,n} = \frac{2\omega^4}{3\pi c^4} \frac{(\sqrt{\varepsilon}-1)^2}{\alpha}(1 - \frac{3}{16\pi})$$

Here, c is the velocity of light, ε is the permittivity of the medium, and ω is the incident wave frequency. The spectral characteristics of systems were calculated in view of the adopted statistical weights $W_{i,n}$. The procedure of forming the systems of clusters provides for the uniform distribution of these formations and is valid at a low concentration of clusters, as a result of which they do not interact with one another. The average value of concentration of clusters of each type in the systems under investigations is 12–13 orders of magnitude lower than the Loschmidt number.

The total dipole moment of cluster $\mathbf{d}_{cl}$ was calculated by the formula

$$\mathbf{d}_{cl}(t) = Z_+^{(k)} \sum_{i=1}^{N_{tot1}} \mathbf{r}_i(t) + Z_-^{(k)} \sum_{j=1}^{N_{tot2}} \mathbf{r}_j(t)$$

where $\mathbf{r}_i(t)$ is the vector indicating the location of nucleus i or point M at the instant of time t; Z is the electric charge located at the center under consideration; index k specifies type of nucleus (or point M); subscript "+" relates to the nucleus carrying a positive electric charge, and "-" – to a negative one; N_{tot1} and N_{tot2} are numbers of positively and negatively charged nucleus in a cluster accordingly.

The permittivity $\varepsilon(\omega)$ as a function of frequency was given by the complex quantity $\varepsilon(\omega) = \varepsilon'(\omega) - i\varepsilon''(\omega)$, which was determined using the equation [34, 35]

$$\frac{\varepsilon(\omega)-1}{\varepsilon-1} = -\int_0^\infty \exp(-i\omega t)\frac{dF}{dt}dt = 1 - i\omega \int_0^\infty \exp(-i\omega t)F(t)dt,$$

where the function $F(t)$ is the normalized autocorrelation function of total dipole moment of cluster,

$$F(t) = \frac{\langle \mathbf{d}_{cl}(t) \cdot \mathbf{d}_{cl}(0) \rangle}{\langle \mathbf{d}_{cl}^2 \rangle}.$$

The cross section of IR radiation absorption is given by

$$\sigma(\omega) = \left(\frac{2}{\varepsilon_v c \hbar \xi}\right) \omega \tanh\left(\frac{\hbar \omega}{2kT}\right) \operatorname{Re} \int_0^\infty dt e^{i\omega t} \langle \mathbf{d}_{cl}(t) \cdot \mathbf{d}_{cl}(0) \rangle, \quad (3)$$

where ε_v is the permittivity of the vacuum, $\hbar = h/2\pi$, h is Planck's constant, and ξ is the refractive index.

The reflection coefficient R is defined as the ratio of average energy flux reflected from the surface to incident flow. In the case of normal incidence of plane monochromatic wave, the reflection coefficient is given by the formula [32]

$$R = \left| \frac{\sqrt{|\varepsilon_1|} - \sqrt{|\varepsilon_2|}}{\sqrt{|\varepsilon_1|} + \sqrt{|\varepsilon_2|}} \right|^2 . \tag{4}$$

Here it is assumed that the wave incidence occurs from a transparent medium (medium 1) to a medium which may be both transparent and nontransparent, i.e., absorbing and scattering (medium 2). The subscripts used with permittivity in expression (4) indicate the medium.

The frequency dispersion of permittivity defines the frequency dependence of dielectric loss $P(\omega)$ in accordance with the expression [33]

$$P = \frac{\varepsilon'' <E^2> \omega}{4\pi},$$

where $<E^2>$ is the average value of square of electric field strength, and ω is the frequency of emitted electromagnetic wave.

Motions with a frequency of less than 1200 cm^{-1} correspond to librations of molecules, and those with a frequency above 1200 cm^{-1} largely describe intramolecular vibrations [36]. Cooperative effects in water clusters result in substantial red-shifts (with respect to water vapor) in both the IR and Raman spectra, often amounting to some hundreds of cm^{-1}. Increased red-shifting of the O-H stretching vibration is common to all systems in which extensive hydrogen bond cooperativity occurs [37-39]. For large water clusters the O-H vibrations are strongly synchronized or phase-locked [38]. At an atomic level, increasingly cooperative hydrogen bonding in large water clusters is associated with an increased negative charge (electrostatic monopole moment) on the oxygen acceptor atom, stabilization of the oxygen atom, and a decrease in both atomic dipole polarization and atomic volume for the hydrogen atom. Increasing cooperativity is more associated with a greater degree of tetrahedral symmetry in water clusters, together with occupation of both pairs of H-donor and lone-pair acceptor sites, than with cluster size alone.

2.3. SPECTRAL CHARACTERISTICS OF CLUSTER SYSTEMS

Investigation of clusterization of the GHG was executed in works [40-47]. Here we shall consider influence of capture of admixture molecules by water clusters on some spectral characteristics of cluster systems. In analyzing frequency-dependent characteristics we restrict ourselves to frequency range $0 \leq \omega \leq 1000$ cm^{-1} because no intramolecular vibrations are present in the model employed by us for systems I-IV. The structure of equimolecular heteroclusters is shown in Figure 1, where configurations of $(CO_2)_{10}(H_2O)_{10}$ and $(CH_4)_{10}(H_2O)_{10}$ clusters are represented. One may observe that both clusters have a significantly irregular structure. Moreover, CO_2 molecules intermix with H_2O molecules and CH_4 molecules gather in one group forming heterocluster's surface. The correlation in the orientation of CO_2 and H_2O molecules can be observed, while CH_4 molecules are disoriented.

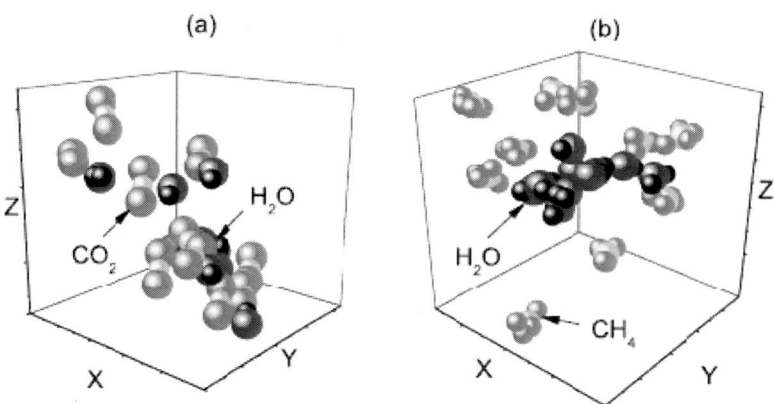

Figure 1. Configurations of clusters: (a) $(CO_2)_{10}(H_2O)_{10}$, (b) $(CH_4)_{10}(H_2O)_{10}$, corresponding to the moment of time 25 ps.

The real ε' and imaginary ε'' parts of dielectric permittivity are not independent. For real positive frequency the connection between them can be described with the help of Cramer - Cronig dispersion relations [32]. The value ε' significantly influences the energy's density of spreading in medium electromagnetic field, and ε'' defines the electromagnetic energy's dissipation in a substance. The calculated values of $\varepsilon'(\omega)$ and $\varepsilon''(\omega)$ for I, II, III and IV cluster systems together with corresponding ε' and ε'' liquid water values obtained by molecular dynamic calculations with TIP4P water model application [48] as well as in the experiment [49] are shown in Figure 2. The dielectric permittivity (both ε' and ε'') of ultra disperse aqueous system containing CH_4 molecules (the system III) is reduced at all frequencies within the range $0 \leq \omega \leq 1000$ cm^{-1} in correlation with corresponding values of "pure" water ultra disperse system (the system I). Moreover, the transparency window can be observed in the frequency range of $640 \leq \omega \leq 790$ cm^{-1} where $\varepsilon''(\omega) = 0$. On the contrary, for systems II and IV, including CO_2 and N_2O molecules, the dielectric permittivity significantly increases, so that its real part (curves 2 and 4, figure 2(a)) at $\omega > 50$ cm^{-1} exceeds the ε' value for liquid water (curve 5, figure 2(a)). One may observe that after the addition of the admixture molecules to $(H_2O)_{10}$ clusters, $\varepsilon'(\omega)$ and $\varepsilon''(\omega)$ the dependences become smoother (with a smaller number of extremes), than corresponding characteristics of system I.

The greenhouse effect caused by atmospheric gases is in fact the absorption of the Earth's thermal radiation by them and a subsequent dissipation of the absorbed energy. The $\sigma(\omega)$ spectrum of Earth's thermal radiation, together with experimental IR radiation absorption spectrum of liquid water [50], is shown in Figure 3(a). The spectrum for water overlaps practically all of the Earth's radiation frequency range and indicates the greatest significance of atmospheric moisture in greenhouse effect creation. The experimental IR spectra of gaseous CO_2 [51], N_2O [52] and CH_4 [53] absorption are represented in Figure 3(b). In the investigated frequency range the locations of the main spectrum peaks for gaseous N_2O and CH_4 coincide. These spectra represent a landmark for location of changes of relevant IR spectra peaks during transition from clusters of pure water to heteroclusters. IR spectra of systems I, II, III and IV presented in Figure 3(c) are calculated by the method described in [43]. The spectra for ultradisperse systems of "pure" water and $(CO_2)_i(H_2O)_{10}$ clusters system have two peaks, the main of which situates at $\omega = 974$ (I) and 960 cm^{-1} (II), and the second peak at 661 cm^{-1} (I) and 724 cm^{-1} (II), it is less expressed. The IR spectra in the frequency range of $0 \leq \omega \leq 1000$ cm^{-1} related with aqueous systems containing N_2O and CH_4 molecules are characterized by one peak. These peaks are situated respectively at the 911 and 340 cm^{-1} frequency. In the observed frequency range the integral intensity of IR radiation absorption by systems II and III significantly decreases. On the contrary, for the system IV this value slightly increases in comparison with the corresponding characteristic of system I.

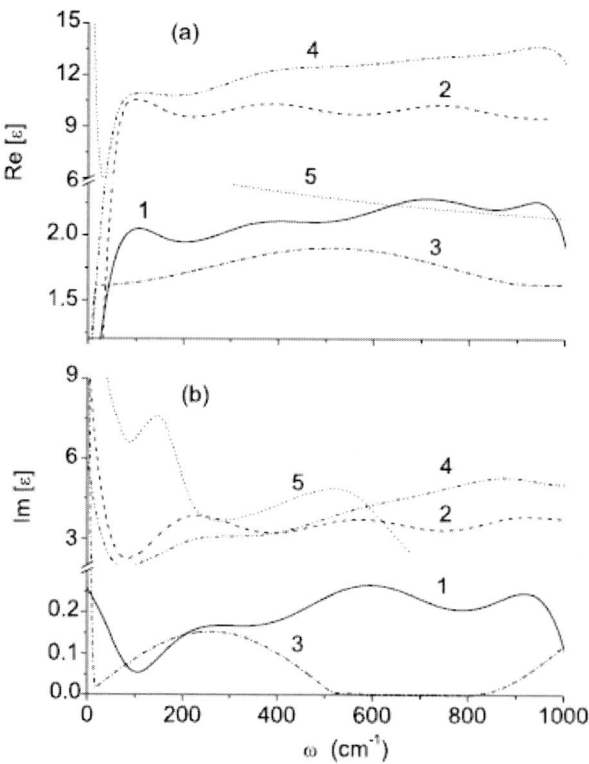

Figure 2. Frequency dependences of (a) real and (b) imaginary parts of the dielectric permittivity for systems: *(1)* $(H_2O)_n$ (I), *(2)* $(CO_2)_i(H_2O)_{10}$ (II), *(3)* $(CH_4)_i(H_2O)_{10}$ (III), *(4)* $(N_2O)_i(H_2O)_{10}$ (IV), *(5)* liquid water: (a) MD with using TIP4P water model [35], (b) experiment [49].

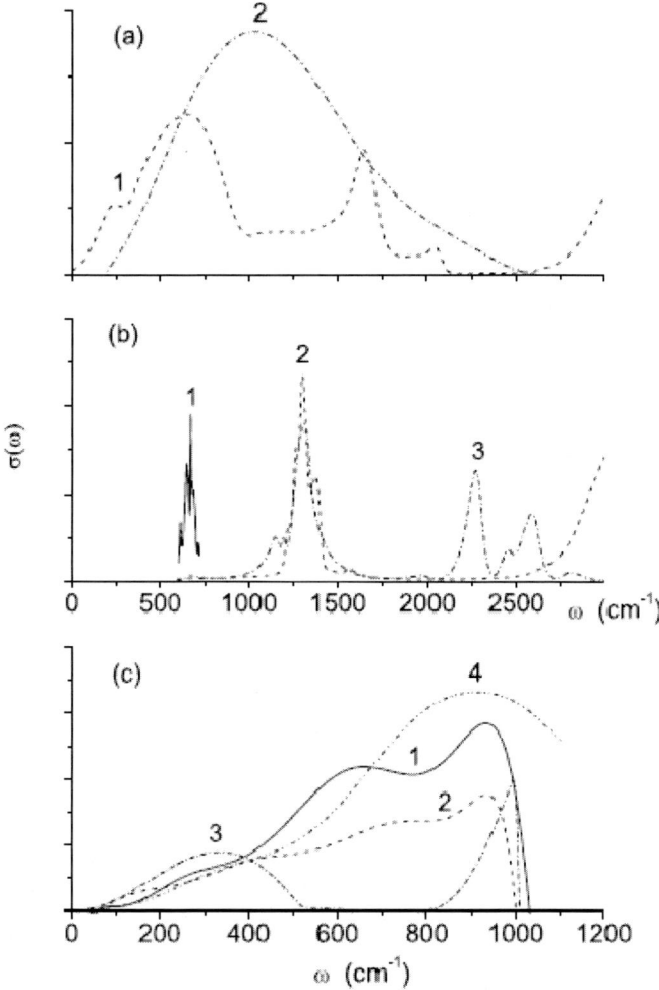

Figure 3. IR absorption spectra, (a) *(1)* experimental spectrum for liquid water [50], *(2)* spectrum of thermal radiation of the Earth at $T = 280$ K; (b) *(1), (2), (3)* experimental spectra for gaseous CO_2 [51], CH_4, and N_2O [52] correspondingly; (c) *(1)* system I, *(2)* II, *(3)* III, *(4)* IV.

Let us consider energy exchange between photons representing falling electromagnetic wave and phonons – collective oscillations of molecules in clusters. The most probable events of this process are [53]:

(1) the absorption of photon with frequency ω (or $\hbar\omega$ energy) with a birth of two phonons with the same ω_1 frequency extending in opposite directions under the law of energy conservation $\hbar\omega = 2\hbar\omega_1$ or $\omega = 2\omega_1$,

(2) the absorption of photon with frequency ω causes appearance of two phonons with different frequencies ω_1 and ω_2, where $\omega = \omega_1 + \omega_2$,

(3) the absorption of photon, the collapse of one phonon and the emergence of another with $\omega = \omega_1 - \omega_2$, $\omega_1 \neq \omega_2$.

The exchange of electromagnetic radiation energy with clusters is an essentially unharmonious process. Due to this the emerged phonon can be characterized not only by one frequency but by a number of frequencies from the appointed interval.

The most probable result of IR radiation interaction with clusters, as well as with crystals [53], is the appearance of two phonons of the same frequency (event 1). We especially refer to this event at $\omega = 974$ cm^{-1} frequency of the main IR spectrum peak of the system I. Then the frequency expected for appearing phonons is defined by $\omega_1 = 487$ cm^{-1} value. The event 2 should be the second one according to the frequency occurrence where absorbed photon energy distributes among excited phonons in uneven portions. To this event we can attribute the emergence of the second IR spectrum peak in the system I at $\omega = 661$ cm^{-1} frequency. The frequency $\omega_2 = \omega - \omega_1 = 174$ cm^{-1} can be thus derived. The third event of the system I, where the frequency of IR spectrum peak localization is defined by $\omega_1 - \omega_2$, is the least probable. In this case the location of expected peak is given by $\omega = 313$ cm^{-1}. In the IR spectrum of the system I in the vicinity of this frequency (at 348 cm^{-1}) there is only one inflection point of dependence $\sigma(\omega)$. It is possible to estimate the influence of admixture molecules on clusters' phonons, and consequently, on corresponding IR spectra by inharmonic contributed by them. In the case of CO_2 molecules the interaction of clusters with IR radiation gives events 1 and 2, and the inharmoniousness is characterized by the quantities $\Delta\omega_1 = \omega_1(II) - \omega_1(I) = -7$ cm^{-1}, $\omega_2(II) - \omega_2(I) = 70$ cm^{-1} (II: $\omega_1 = 480$ cm^{-1} and $\omega_2 = 280$ cm^{-1}). The event 1 with phonon frequency $\omega_1 = 455.5$ cm^{-1} takes place at interaction of IR radiation with $(N_2O)_i (H_2O)_{10}$ clusters, and the inharmoniousness is defined by the value $\Delta\omega_1(IV) = 31.5$ cm^{-1}. From the minimum unharmonious effect estimation it follows that for $(CH_4)_i$ $(H_2O)_{10}$ clusters the realization of event 3 with phonon frequencies $\omega_1 = 650$ cm^{-1} and $\omega_2 = 310$ cm^{-1} leading to a quantity of inharmoniousness $\Delta\omega_2$ (III) = 136 cm^{-1}, is more probable. Thus, according to the quantity of contributed inharmoniousness (from greater to smaller) the admixture molecules locate as CH_4, CO_2, N_2O.

The frequency distribution of power dissipated by cluster systems under consideration is given in Figure 4. One can see that the addition of CO_2, CH_4 and N_2O molecules to water clusters causes a significant increase in the rate of energy dissipation. The energy of absorbed IR radiation is dissipated most rapidly by system II with the maximum of dissipation at a frequency 800 cm^{-1}. The system IV has next rate of energy dissipation with a maximum of spectrum $P(\omega)$ at frequency 1036 cm^{-1}. The system of water clusters with absorbed CH_4 also strengthens IR radiation emission power. Water clusters exhibit the highest rate of dissipation of stored energy at frequency ω = 974 cm^{-1}, and clusters which absorbed CH_4 molecules (system III) – at 1014 cm^{-1}.

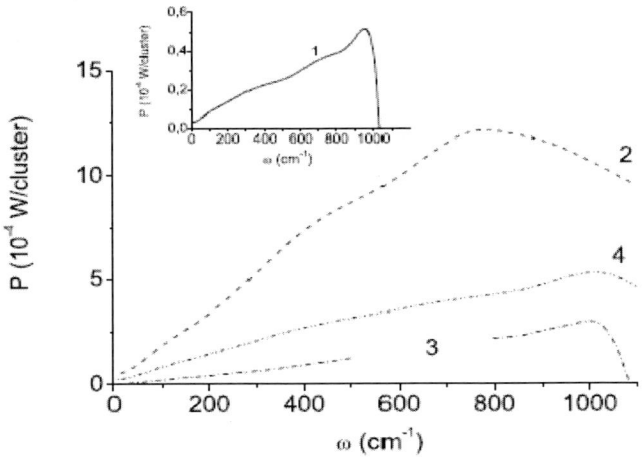

Figure 4. The frequency dependence of dielectric loss for systems: *(1)* I (in the inset), *(2)* II, *(3)* III, *(4)* IV.

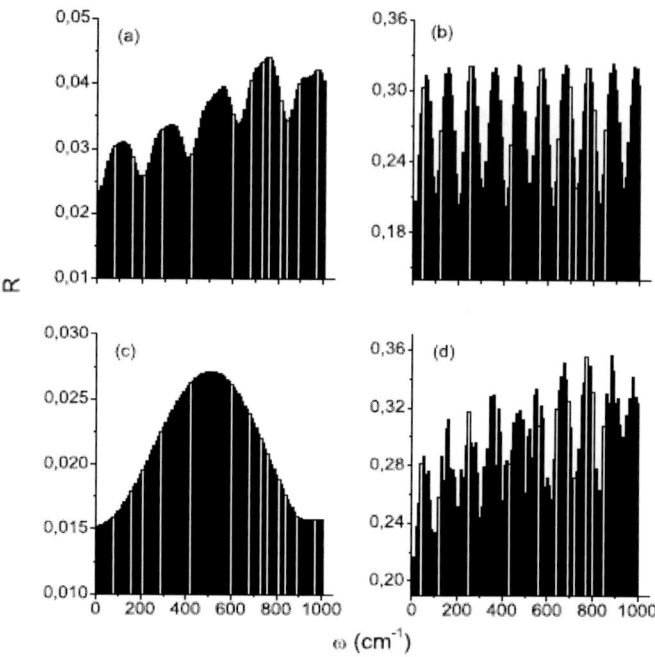

Figure 5. Reflection coefficient for cluster systems: (a) $(H_2O)_n$, (b) $(CO_2)_i(H_2O)_{10}$, (c) $(CH_4)_i(H_2O)_{10}$, (d) $(N_2O)_i(H_2O)_{10}$.

The R reflection coefficients of a flat monochromatic wave for pure water clusters and water clusters which absorbed CO_2, CH_4 and N_2O molecules are shown in Figure 5. Water clusters which have absorbed linear CO_2 and N_2O molecules have the high R values. The presence of big (ten) number of peaks points to the orientation orderliness of admixture molecules in these clusters. At the absorption of linear not polar molecules (CO_2) peaks of R spectra have approximately the identical intensity and when the similar polar molecules (N_2O) are absorbed the intensities of peaks considerably differ. The amplification of IR radiation reflection by water clusters that absorbed CO_2 and N_2O molecules is mainly caused by the compaction of water aggregates due to linear molecules' penetration [47]. Adsorption of spherically symmetric nonpolar CH_4 molecules smoothes peaks in R distribution of pure water clusters system making it unimodal. It results in some reduction of the reflection intensity. Methane molecules do not penetrate inside of clusters, they are adsorbed on a surface creating a high roughness and by that reducing the R coefficient.

Chapter 3

CLUSTER MECHANISM OF OZONE DESTRUCTION BY CHLORINE AND BROMINE IONS

In the chemistry of atmospheric particles, halogen activation, which describes the liberation of halogen-containing components from sea salt particles, is considered. Subsequently, these components undergo photolysis to release exceedingly reactive Cl and Br atoms. Photochemical reactions in clouds occur in the liquid phase with the participation of OH and HO_2 radicals and other oxidizing components (H_2O_2, O_3) supplied from the gas phase. Water clusters are present in clouds and can also exist in a cloudless atmosphere [54]. Part of molecules in the atmosphere experience ionization under the action of cosmic rays and radiation of radioactive rocks and radon in air itself. The ionization of molecules releases electrons. Chlorine and bromine atoms easily attract free electrons to produce negatively charged ions. Chlorine or bromine ions interact with ozone and split it into oxygen molecules and atoms (possibly, into three oxygen atoms). This reaction is facilitated in the presence of water. In other words, the reaction has the highest rate on the surface of water clusters or drops. Fast chlorine or bromine ions first reach the surface of water clusters. Destruction of ozone on a surface of water clusters at the presence of chlorine ions is investigated in works [55-57].

3.1. METHOD

Here we consider flexible models of molecules. Flexibility was created using the procedure developed in terms of Hamilton dynamics in [58, 59]. Let us consider a diatomic molecule, and let atoms a and b in it be separated by the distance

$$q = \|\mathbf{r}_a - \mathbf{r}_b\|,$$

where $\mathbf{r}_a$ and $\mathbf{r}_b$ are the vectors that determine the positions of atoms. Let $\mathbf{v}_a$ and $\mathbf{v}_b$ be the corresponding velocities, and the reduced mass have the form

$$\mu = \frac{m_a m_b}{m_a + m_b}.$$

The size of the molecule consisting of atoms a and b is determined by the equality of the total potential force $\mathbf{f}(\mathbf{q}) = -\frac{\partial \mathbf{r}}{\partial \mathbf{q}} \nabla \Phi(\mathbf{r})$ to the centrifugal force $-\mu q \omega^2$, that is,

$$-\mu q \omega^2 - \mathbf{f}(\mathbf{r}) \frac{\partial \mathbf{r}}{\partial \mathbf{q}} = 0,$$

where $\omega = \|\mathbf{v}_a - \mathbf{v}_b\|/q$ is the angular velocity. The minimization of contributions to the potential energy U of each generalized coordinate yields

$$\frac{\partial}{\partial q_i} H(\mathbf{r}, \mathbf{v}) = \frac{\partial}{\partial q_i} \left(\frac{1}{2} \mu_i q_i^2 \omega_i^2 + U(\mathbf{r}) \right) = 0.$$

This method is applicable to molecules with arbitrary compositions [60]. The use of flexible model of molecules allows expanding the range of frequencies under study.

The dynamics of a system of molecules was reproduced by water–water intermolecular interaction potential [19] and oxygen–oxygen and oxygen–water interactions described by an atom–atom potential calculated in the Gordon-Kim approximation with the application of spherical average of electronic densities [25, 26]. The Coulomb interaction of the Cl^- and Br^- ions with water is determined by their electric charge = $-1e$, where e is the elementary charge unit. The Ion–H_2O non-Coulomb interaction is included as Lennard-Jones $Cl^- - O$ or $Br^- - O$ interaction with the parameters from [61, 62]. In addition to Coulomb interaction, we also considered atom–atom interaction between Cl^- or Br^- ions described by (2) with parameters from [26]. The $q_+ = 0.19e$ positive charge was set on the central atom of O_3 molecule and the $q_- = -0.095e$ on each of its side atoms [62]. The O_3 molecule was electrically neutral as a whole. Atoms in O_2 molecule had no electric charges.

The simulation of the interaction of $(Cl^-)_i(H_2O)_{50-i}$ or $(Br^-)_i(H_2O)_{50-i}$ clusters with a gas medium containing i H_2O molecules and six O_3 or O_2 molecules began with creating the configuration of the equilibrium cluster of water $(H_2O)_{50}$ with the kinetic energy corresponding to 250 K in molecular dynamics calculations. The positions of chlorine or bromine ions in the cluster were determined using the system of coordinates fixed at the center of mass of the cluster. The cluster was approximated by a sphere, and water molecules situated most closely to the emergence of the axes of coordinates from this sphere were found. These molecules (their number was from one to six) were displaced outside along the coordinate axes through distances of 0.6–0.7 nm from the former positions of their centers of mass, into which Cl^- or Br^- ions were introduced. The number of ions equaled the number of displaced molecules. The interaction of the newly formed $(Cl^-)_i(H_2O)_{50-i}$ or $(Br^-)_i(H_2O)_{50-i}$ cluster with water and ozone or oxygen molecules was studied in an ensemble with an external thermostat [63], whose temperature was 250 K. The cutoff radius for intermolecular interactions was 0.9 nm in our model. Six O_3 or O_2 molecules were situated predominantly on the side of the lower cluster part. Fairly compact arrangement of these molecules facilitated studies of their influence on the behavior of Cl^- or Br^- ions.

The Raman and infrared spectra of clusters can be calculated via the autocorrelation functions of polarizability and dipole moment, respectively. A polar molecule is characterized by a permanent (gas-phase) dipole moment $\mathbf{d}_{i,0}$ and polarizability tensor $\boldsymbol{\alpha}^P_{i,0}$. Interaction with neighboring molecules creates induced dipole moment and polarizability of molecule i. Each model molecule can be treated as a polarizable point dipole situated in the center of mass of the molecule. The $\mathbf{d}_i$ dipole moment of molecule i and its polarizability $\boldsymbol{\alpha}^P_i$ are related because of molecule interactions with the environment [64],

$$\mathbf{d}_i = \mathbf{d}_{i,0} + \boldsymbol{\alpha}^P_{i,0} \sum_{j \neq i} \mathbf{T}_{ij} \mathbf{d}_j, \tag{5}$$

$$\boldsymbol{\alpha}^P_i = \boldsymbol{\alpha}^P_{i,0} + \boldsymbol{\alpha}^P_{i,0} \sum_{j \neq i} \mathbf{T}_{ij} \boldsymbol{\alpha}^P_j. \tag{6}$$

We used the anisotropic gas-phase polarizability tensor $\alpha^P_{xx,yy,zz}$ = {1.495, 1.626, 1.286} Å^3 for the water molecule [64]. Ozone and oxygen molecules were characterized by the isotropic experimental polarizability values 2.85 and 1.57 Å^3 [65], respectively. Equations (5) and (6) for $\mathbf{d}_i$ and $\boldsymbol{\alpha}^P_i$ were solved by inverting the matrix and using the $\mathbf{d}_{i,0}$ and $\boldsymbol{\alpha}^P_{i,0}$ values on the right-hand side.

The cross section of IR radiation absorption is given by equation (3). For depolarized light, the Raman spectrum is given by the equation [64]

$$J(\omega) = \frac{\omega}{(\omega_L - \omega)^4}\left(1 - e^{-\hbar\omega/kT}\right) \operatorname{Re} \int_0^\infty dt e^{i\omega t} \langle \Pi_{xz}(t)\Pi_{xz}(0)\rangle, \tag{7}$$

where

$$\Pi(t) \equiv \sum_{j=1}^{N}\left[\alpha_j^p(t)-\left\langle\alpha_j^p\right\rangle\right],$$

ω_L is the exciting laser frequency, Π_{xz} is the xz component of $\Pi(t)$, the x axis is directed along the molecular dipole, and xy is the molecular plane. Simulations were performed for $\omega_L = 19436.3$ cm^{-1} (the argon laser green line, $\lambda = 514.5$ nm).

In liquid water, the interaction of Cl$^-$ and Br$^-$ ions with H_2O molecules is limited to the electrostatic attraction of water dipoles to anions. In a small water cluster, Cl$^-$ and Br$^-$ ions experience mutual repulsion and cannot be held fairly long close to water molecules. We performed calculations for six systems $(Cl^-)_i(H_2O)_{50-i} + i\, H_2O + 6O_2$, $i = 1, \ldots, 6$ and six systems $(Cl^-)_i(H_2O)_{50-i} + i\, H_2O + 6O_3$. The same calculations were carried out for systems containing ions of bromine. Systems with $i = 1-6$ are denoted by I–VI in the presence of oxygen and by VII–XII in the presence of ozone. We shall put one stroke at number of the system containing ions of chlorine, and two strokes - at the system number including ions of bromine. The criterion of the addition of O_3 or O_2 molecules to a water cluster was the establishment of $r_{O-O} \leq 0.35$ nm distances.

3.2. THE SPECTRAL CHARACTERISTICS OF CHLORINE-CONTAINING WATER CLUSTERS IN THE PRESENCE OF OZONE AND OXYGEN

Because of Coulomb repulsion from negatively charged oxygen ions and mutual repulsion, the Cl$^-$ ions in $(Cl^-)_i(H_2O)_{50-i}$ clusters rushed outside. The ions leaving a cluster pushed water molecules apart. This and interaction with environment molecules (H_2O and O_3 or O_2) heated the $(Cl^-)_i(H_2O)_{50-i} + i\, H_2O + 6O_3$ ($6O_2$) system. The strongest heating to 311 K (at time ~3 ps) was observed when the cluster of water with six chlorine atoms was surrounded by ozone molecules. The algorithm that we used allowed of the decomposition of O_3 molecules with energies of 1.3 eV or higher. This energy is sufficient for splitting an ozone molecule surrounded by water molecules. However, in reality, there was no decomposition of ozone. As a rule, in 2 ps, that is, by the time $t = 5$ ps, the system was cooled because of velocity scaling and reached its initial temperature ~250 K. By the time 1.2 ps, two Cl^- ions left the $(Cl^-)_6(H_2O)_{44}$ cluster, other two ions directly interacted with O_3 molecules (one of these ions interacted with two O_3 molecules at once), one ion was "bound" to a water molecule, and one ion was almost free (Figure 6). By the time 2.8 ps, all Cl^- ions left the cluster, and O_3 molecules, conversely, approached it. Clusters with four and two Cl^- ions lost these ions in 2.8 and 2.2 ps, respectively. A similar interaction picture was observed when oxygen rather than ozone molecules approached the cluster. The system was then free of ions in 2.8 ($6\,Cl^-$), 2.4 ($4\,Cl^-$), and 2.2 ps ($2\,Cl^-$).

The IR absorption spectra are determined by the behavior of the autocorrelation function of the total dipole moment $\mathbf{d}_{cl}$ of a cluster. When oxygen molecules without permanent dipole moments are added to a water cluster, they weakly influence the $\mathbf{d}_{cl}$ value determined by the aqueous component of the cluster. For this reason, the perturbation force introduced by Cl^- ions is approximately proportional to the number of ions N_{Cl^-}. The integral intensity I_{tot} of IR radiation absorption spectra therefore grows as N_{Cl^-} increases (Figure 7a). The opposite situation arises if ozone molecules are absorbed. When ozone molecules, whose permanent dipole moment is lower than that of water (0.12 D), are added to a cluster, they create additional "beats" during the evolution of the $\mathbf{d}_{cl}$ value. The action of Cl^- ions on the cluster strengthens these oscillations with different frequencies and phases. The larger the number of Cl^- ions in the cluster, the stronger are beats created by O_3 molecules. As a result, the ability of the system to absorb IR radiation decreases (the I_{tot} value decreases) as the number of ions increases (Figure 7b). The I_{tot} value for systems $I'-VI'$ increases nonmonotonically and that for systems $VII'-XII'$ decreases as the number of Cl^- ions grows. The I_{tot}

intensities are related as 1 : 0.80 : 1.22 : 1.10 : 2.83 : 1.83 for systems $I'-VI'$ and as 1 : 1.04 : 0.86 : 0.79 : 0.99 : 0.94 for systems $VII'-XII'$. In all cases, the absorption peak is in the vicinity of the frequency $\omega = 960 \pm 20$ cm^{-1}. The main absorption of liquid water is observed at 690 cm^{-1} [50], and that of gas ozone-oxygen mixtures, in the vicinity of 996 cm^{-1} [66].

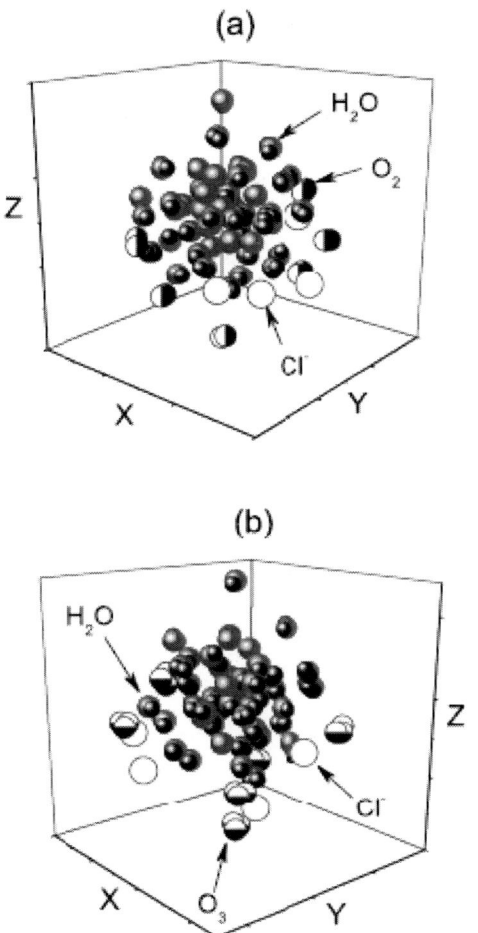

Figure 6. Configurations of systems: $(Cl^-)_6(H_2O)_{44} + 6H_2O + 6X$, corresponding to the moment of time 1.2 ps: (a) – $X = O_2$, (b) – $X = O_3$.

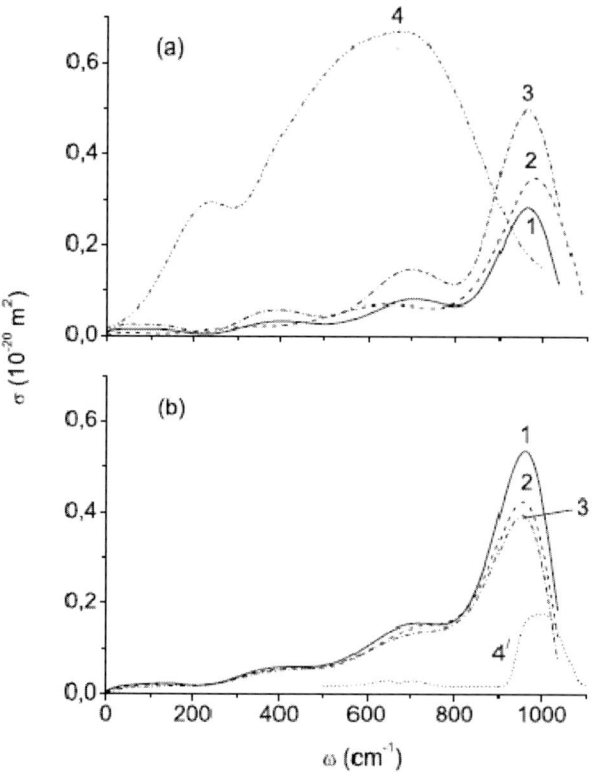

Figure 7. IR absorption spectra of systems with (a) oxygen and (b) ozone: (*1*) systems I' and VII', (*2*) III' and IX', (*3*) VI' and XII', (*4*) $\sigma(\omega)$ function of bulk liquid water, experimental [50], and (*4'*) experimental spectrum of an $O_2 + O_3$ mixture of gases [66].

The Raman spectra of aqueous systems containing oxygen or ozone are still more different than the IR spectra (Figure 8). The Rayleigh line at $\omega = 0$ is excluded from the spectrum. The positions of the first peaks coincide (26 cm^{-1}), but there is no correspondence between the other peaks. In addition, the integral intensity of Raman spectra is noticeably higher with oxygen than with ozone in aqueous systems. The Raman spectrum for liquid water is characterized by low-frequency peaks at and -30 cm^{-1}, as well as by a peak near 170 cm^{-1} [67]. The peak at 60 cm^{-1} is formed by the bending of hydrogen bonds between water molecules; the peak at 170 cm^{-1} appears due to bond stretching [64]. The position of the newly formed last peak at 950 cm^{-1} in the spectrum of the system with six oxygen molecules and six chlorine ions is close to the position of the peak in the Raman spectrum of an aqueous solution of chloride dioxide (945 cm^{-1}) determined experimentally [68]. The peak observed in femtosecond measurements of a solution of OClO in water was assigned to a fundamental transition to a new energy level with symmetrical dilatation of the OClO molecule.

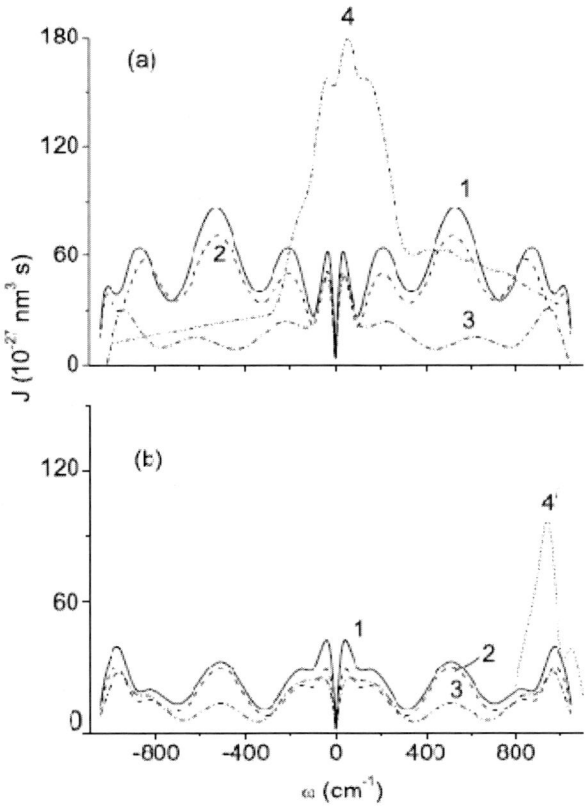

Figure 8. Raman spectra of systems with (a) oxygen and (b) ozone: (*1*) systems I' and VII', (*2*) III' and IX', (*3*) VI' and XII', (*4*) liquid water, experimental [67], and (*4'*) solution of OClO in water, experimental [68].

A study of the photolysis of OClO in aqueous solution showed the possibility of separation of dielectric relaxation related to solvent (water) libration modes and mechanical solvation caused by low-frequency translational motion [69]. A mechanical response of a solvent is usually provided by its one or two molecules in the first solvation shell. According to the computer simulation results, 80% of water solvation that follows OClO photoexcitation occurred during a 2×10^{-14} s time interval, whereas the remaining 20% of the response were diffusion in nature and could be observed on a picosecond time scale [69].

Two characteristic times can also be identified for excitation related to the removal of Cl^- ions from a water cluster. A short time interval (~3 ps) corresponds to the excitation or interaction of ions with water and ozone (oxygen) molecules. The dielectric relaxation interval is ~10 ps. During this time, the cluster "forgets" the value and direction of its dipole moment at time $t = 0$. The dissociation of the free O_3 molecule (in the vacuum) usually occurs when energy of 4–5.5 eV is imparted to it. Chlorine ions flying out of the cluster do not possess such energy. For this reason, we did not observe the decomposition of ozone molecules.

As the strength and duration of perturbation caused by the "bombardment" of molecules (H_2O and O_3 or O_2) by Cl^- ions increased, the integral intensity of the IR absorption spectrum decreased in the presence of ozone and increased with the addition of oxygen. The main absorption frequency remained constant. The Raman spectra substantially changed their shape and intensity when absorbed oxygen was replaced with ozone. Irrespective of the number of Cl^- ions initially contained in a water cluster and molecules it absorbs (O_3 or O_2), the main Raman spectrum frequency remains unchanged. The absorption of ozone causes more rapid decrease in the intensity of Raman spectra compared with the absorption of oxygen.

IR emission spectra $P(\omega)$ for oxygen-containing systems are characterized by the presence of a principal band in the $800 \leq \omega \leq 1000$ cm^{-1} frequency range (Figure 9a). The integral intensity I_{rad} of emission spectra of systems under consideration increases significantly in the presence of Cl^- ions. The ratio of I_{rad} values for the $P(\omega)$ spectra of $(O_2)_6(H_2O)_{50}$ cluster and systems II′, IV′ and VI′ is equal to 1 : 1.45 : 1.09 : 1.70. The intensity of IR radiation emission by the $(O_3)_6(H_2O)_{50}$ cluster increases as a result of the action of chlorine ions on it (Figure 9b). The ratio of integrated intensities I_{rad} in the $P(\omega)$ spectra of $(O_3)_6(H_2O)_{50}$ cluster and systems VIII′, X′, and XII′ is 1:1.34:1.16:1.07. At the presence of Cl^- ions displacement of the main maximum of $P(\omega)$ spectrum in the area of low frequencies occurs more strongly for systems containing oxygen. The increase in the number of Cl^- ions acting on the cluster results in a better resolution of the second and third peaks of the $P(\omega)$ spectrum.

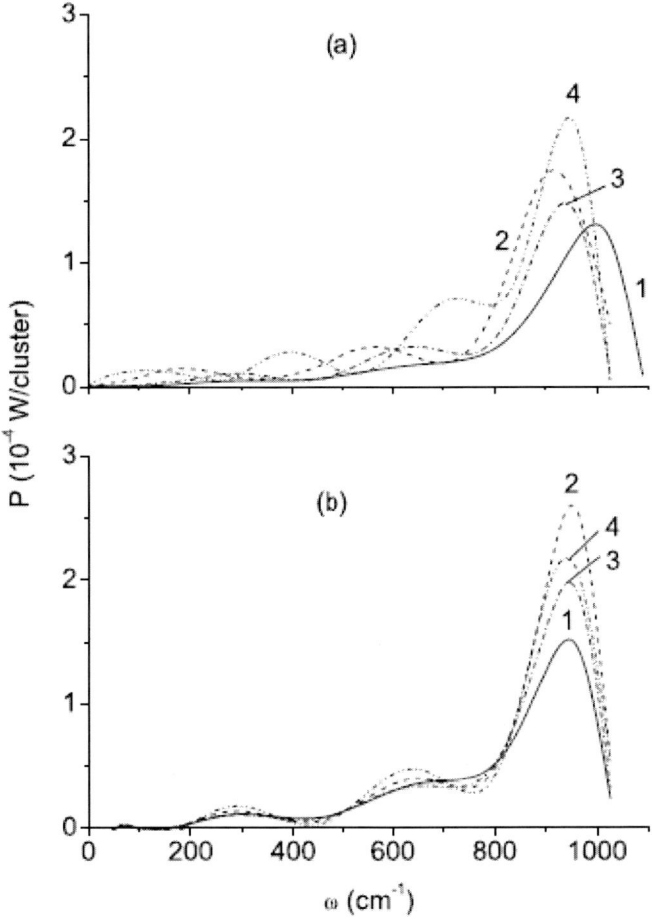

Figure 9. IR emission spectra of systems $(Cl^-)_i (H_2O)_{50-i} + i\, H_2O + 6\,X$: (a) $X = O_2$, (b) $X = O_3$. (1) $i = 2$, (2) $i = 4$, (3) $i = 6$; (4) $P(\omega)$ function of the $(O_3)_6 (H_2O)_{50}$ cluster.

3.3. Dielectric Properties of Bromine-Containing Water Clusters in the Presence of Ozone and Oxygen

Difference in behaviour of Br^- ions in water cluster, interacting with O_2 or O_3 molecules, can appear because of their greater weight than that of Cl^- ions ($m_{Br^-}/m_{Cl^-} = 2.25$). Configurations of the $6Br^-(H_2O)_{44} + 6H_2O + 6X$ cluster systems, received to the end of calculation, i.e. to the moment of time 25 ps, are shown in figure 10. Here O_2 molecules (Figure 10a) or O_3 ones (figure 10b) under X molecules are meant. In both cases six free water molecules have joined to $6Br^-(H_2O)_{44}$ cluster. Three of six Br^- ions have abandoned the cluster at presence in system of oxygen. One of these ions is depicted in the bottom part of figure. Three O_2 molecules have evaporated also together with the left ions while three other molecules of oxygen have joined to the cluster.

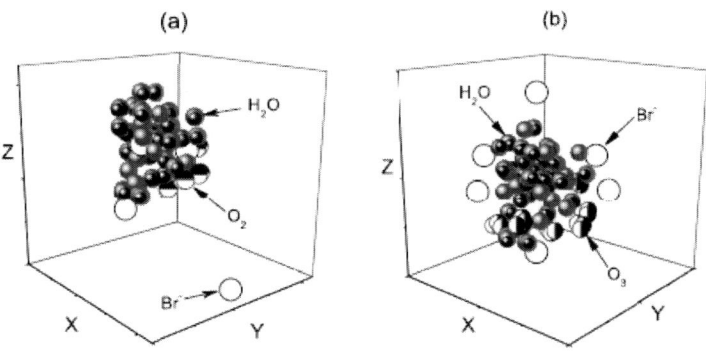

Figure 10. Configurations of systems: $(Br^-)_6(H_2O)_{44} + 6H_2O + 6X$, corresponding to the moment of time 25 ps: (a) $X = O_2$, (b) $X = O_3$.

The effect from the presence of molecules of the ozone made another result. All six O_3 molecules were adsorbed by cluster. Thus six Br^- ions have left on the cluster surface, but they have not left from cluster so far as we could consider them as evaporated ones. We consider ions or molecules, for which non-coulomb interaction with other molecules vanishes because of cutting the appropriate interaction potential at distance 0.9 nm, as evaporated ones. Such ions or molecules did not come back to the cluster.

Frequencies and intensities of normal vibrations are defined by weight and a kind of atoms, bonding force and spatial arrangement of atoms (lengths and corners of the bond). Absorbed ions, changing the cluster structure, influence intensity of the IR spectrum and a smaller degree on intensity of Raman spectrum.

Absorption infra-red $\sigma(\omega)$ spectra of some systems containing bromine are depicted in Figure 11. The addition of six oxygen molecules to the $(H_2O)_{50}$ cluster, results in an increase of the integrated I_{tot} intensity of IR radiation absorption by factor 2.3. However, if a water cluster preliminary absorbed Br^- ions (at equivalent desorption of water molecules) then such amplification of I_{tot} value is not observed, and the I_{tot} magnitude at presence of four and more Br^- ions and absorption of oxygen, decreases. The position of the main peak of the IR spectrum of disperse water system with molecular oxygen does not change, and the second in intensity peak is deformed in a shoulder. The location of the basic peak is shifted (on ~ 100 cm^{-1}) in area of low frequencies concerning position of the main peak of an experimental IR spectrum of liquid bulk water [50]. Localization of minimum between two peaks of $\sigma(\omega)$ spectra of cluster systems falls to an interval of the maximal intensity of an experimental IR spectrum of gaseous bromine hydride [52], and fast rise of spectrum intensity of systems under consideration corresponds to the position of the main maximum of an experimental IR spectrum of gaseous $O_2 + O_3$ mix [66].

Opposite transformations of $\sigma(\omega)$ spectrum occur at a change of number of Br^- ions in a cluster at the presence of ozone in the system. The addition of O_3 molecules to cluster gives reduction of I_{tot} magnitude by factor 7.3. This decrease is less expressed with the growth of the number of Br^- ions in a cluster. In other words, adsorption of ozone goes with the increase of I_{tot} values if water cluster consistently attaches Br^- ions. So at presence of two Br^- ions the intensity of IR spectra at adsorption of ozone decreases by factor 2.1 and when 4 Br^- ions are in cluster the I_{tot} magnitude is reduced only by factor 1.07. The increase of intensity of $\sigma(\omega)$ spectra stops, when the number of Br^- ions exceeds 4.

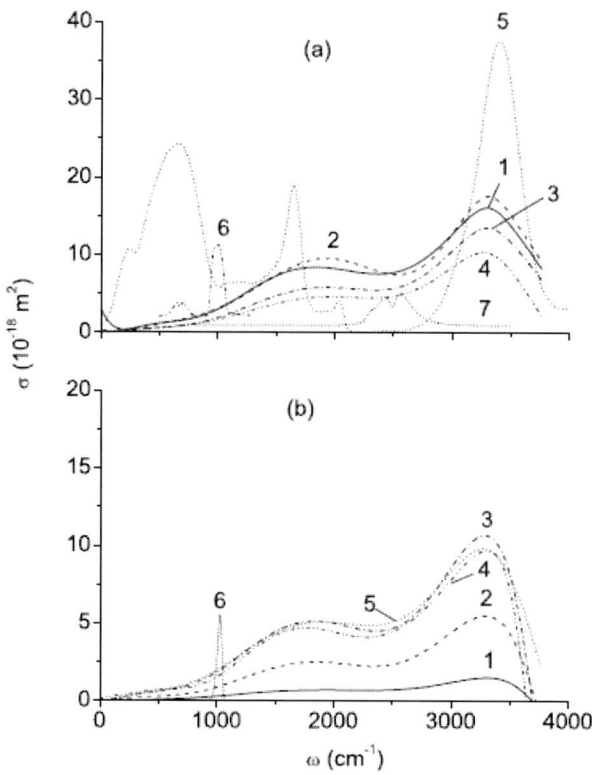

Figure 11. IR absorption spectra for systems $(Br^-)_i (H_2O)_{50-i} + i\, H_2O + 6X$: (a) $X = O_2$, (b) $X = O_3$. (1) $i = 0$, (2) $i = 2$, (3) $i = 4$, (4) $i = 6$; (a): (5) $\sigma(\omega)$ function of of bulk liquid water, experiment [50], (6) experimental spectrum of an $O_2 + O_3$ mixture of gases [66], (7) $\sigma(\omega)$ spectrum of gaseous HBr, experiment [52]; (b): (5) function $\sigma(\omega)$ of the $(H_2O)_{50}$ cluster, (6) experimental spectrum of gaseous O_3 [70].

The position of the main peak of IR spectrum practically does not change and approximately corresponds to the frequency 3300 cm^{-1} during the change of the number of Br^- ions in a cluster. Here (as well as at presence of oxygen in the system) red shift of position of the main maximum of a spectrum concerning localization of the basic peak of an experimental IR spectrum of liquid water is observed. The minimum dividing peaks of $\sigma(\omega)$ spectra of cluster systems falls to an area of the greatest intensity of the IR spectrum of gaseous bromine hydride. The fast increase of intensity of $\sigma(\omega)$ spectra of these systems occurs in frequency range where the peak of IR spectrum of gaseous ozone [70] is located.

Nonpolar O_2 molecules create effect "dilution" owing to the decrease of the total dipole moment magnitude. Br^- ions perturb an internal electric field of cluster, accelerating attenuation of autocorrelation function of d_{cl} magnitude. As a result intensity of the IR spectrum decreases with growth of number of bromine ions. The grouped polar O_3 molecules create poorly varying electric field which orders the dipole moments of water molecules. As a result the d_{cl} magnitude is essentially increased, and intensity of the IR spectrum grows in the process of increase of the number of Br^- ions in the system. Amplification of perturbation due to addition to the cluster of over four Br^- ions results in faster attenuation of autocorrelation function of d_{cl} dipole moment. As a consequence, evolved at $i = 4$ the intensity of IR spectrum of aqueous-ozone system starts to reduce.

Raman $J(\omega)$ spectra for bromine-containing systems (I'', VI'', VII'', and XII'') are shown in Figure 12. Adsorption of six oxygen molecules results in an increase of integrated J_{tot} intensity of Raman spectrum of $(H_2O)_{50}$ cluster by the factor 1.6. At the presence of oxygen the $J(\omega)$ spectrum poorly changes the intensity and the shape at variation of number of Br^- ions in a cluster. So at absorption of 6 Br^- ions by aqueous-oxygen system the J_{tot} magnitude has decreased only by 7.1 %. In the systems containing up to six Br^- ions, red shift of the basic peak on quantity ~ 110 cm^{-1} concerning a position of the main peak of experimental Raman spectrum of liquid water [71] is observed. This peak has blue shift on quantity ~ 130 cm^{-1} concerning localization of the expressed peak in the appropriate experimental spectrum of the Br_2 clathrate hydrates [72]. Addition of six molecules of ozone results in easing integrated J_{tot} intensity of Raman spectrum for

$(H_2O)_{50}$ cluster by factor 1.5. The quantity of Br^- ions taking place in cluster is essentially reflected in the shape of $J(\omega)$ spectra of the cluster systems containing ozone. $J(\omega)$ spectra considerably reduce the intensity in the process of increase of the Br^- ions number. The J_{tot} magnitude of aqueous-ozone system at the presence of 6 Br^- ions is lower by factor 4.3, than J_{tot} value of such system at absence of bromine ions.

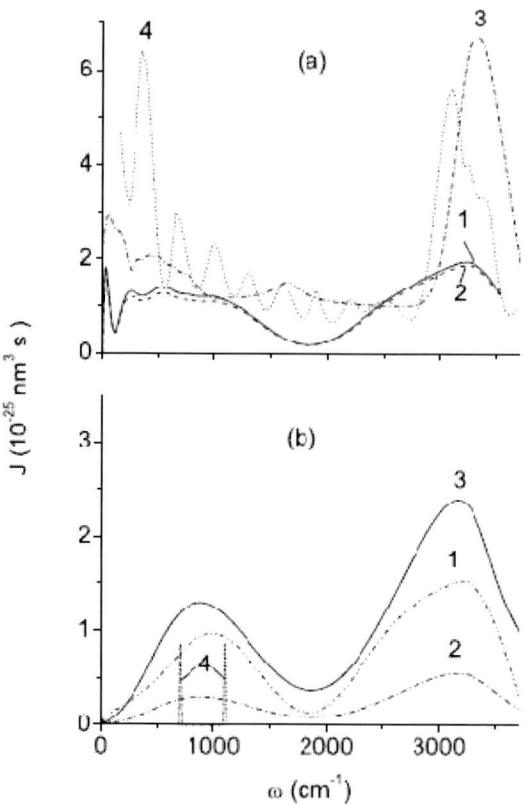

Figure 12. Raman spectra: (1–2) the $(Br^-)_i(H_2O)_{50-i} + i\, H_2O + 6\,X$ systems; (a) $X = O_2$, (b) $X = O_3$; (1) $i = 0$, (2) $i = 6$; (a): (3) liquid water at 293 K, experiment [71], (4) the Br_2 clathrate hydrates at 266 K, experiment [72]; (b): (3) the $(H_2O)_{50}$ cluster, (4) gaseous O_3, experiment [73].

Red and blue (about 170 and 20 cm^{-1}) shifts of maximum position of the Raman spectrum of cluster systems concerning the appropriate peaks of experimental spectra of water and bromine clathrate hydrates are observed. The position of first (less intensive) peak of $J(\omega)$ spectra for cluster systems containing ozone gets on an interval dividing two peaks of the appropriate experimental spectrum of gaseous ozone [73].

Average $\overline{\alpha}^p$ polarizability per molecule goes down in comparison with the designed α^p magnitude for water cluster at addition of nonpolar molecules having lower polarizability α^p (0.793 Å^3) than α^p magnitude (1.49 Å^3) for molecules of liquid water. Presence of Br^- ions renders weak influence on behaviour of autocorrelation function of fluctuations of the polarizability. As a result we have close Raman spectra of the investigated systems containing oxygen. Other situation arises for the systems containing ozone. Polar molecules of ozone have higher polarizability (α^p = 2.7 Å^3), than water molecules. Therefore the $\overline{\alpha}^p$ magnitude here is higher, than α^p for water cluster. Disordering of the dipole moments amplifies with increase of number of Br^- ions and values of polarizability are smoothly reduced. Here fluctuations of α^p magnitude are lower, than that for systems with oxygen. As a result intensity of $J(\omega)$ spectrum for the systems containing ozone is reduced in process of increase of number of Br^- ions.

Adsorption of both molecules of oxygen and ozone by the $(H_2O)_{50}$ cluster strengthens integrated P_{tot} intensity of emission spectra of IR radiation. And, as a result of addition of six O_2 molecules the P_{tot} magnitude has increased by factor 3.8, and at addition of six O_3 molecules to the cluster – by factor 4.1 (Figure 13). Emission spectra of cluster systems essentially change the intensity at a variation of number of Br^- ions in cluster. Thus, as a rule, intensity of emission spectra is reduced in case of presence the oxygen in system and grows, when clusters adsorb ozone. For example, at presence of 6 Br^- ions the $P(\omega)$ spectrum reduces integrated intensity by factor 1.1 at adsorption of oxygen and increases it by factor 6.4 when ozone joins. The position of the main peak of emission spectrum on the average is a little bit displaced (by ~ 20 cm^{-1}) aside high frequencies at addition of bromine ions in oxygen-containing system and is not strong (up to ~ 35 cm^{-1}) to be shifted aside low frequencies for the system containing ozone. The maximal reduction of intensity of emission spectra for systems with oxygen occurs at presence of four Br^- ions in cluster, and the maximal increase of this characteristic is observed for systems with ozone at the presence of six Br^- ions in the aggregate.

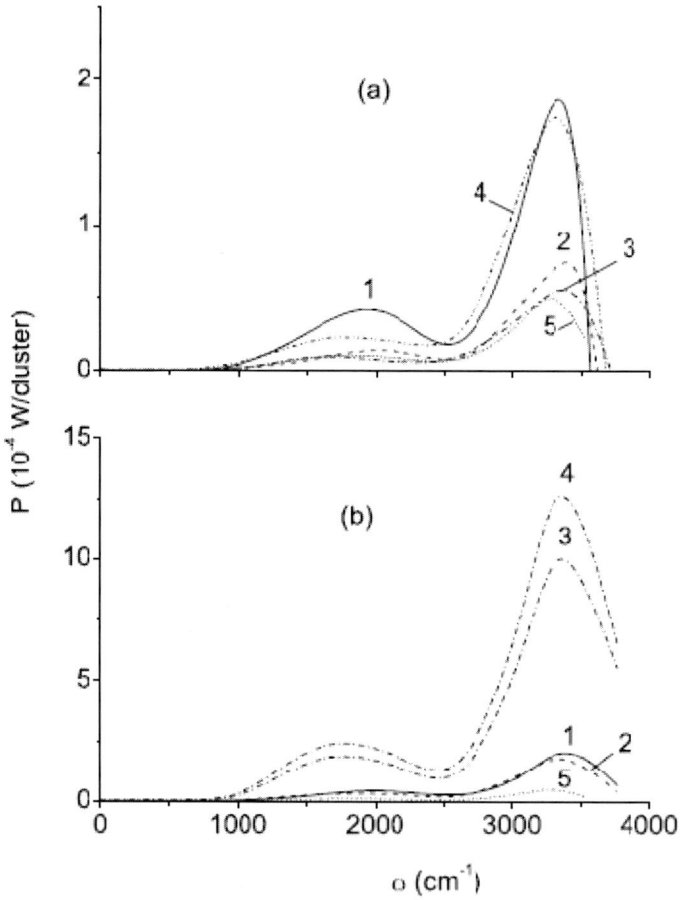

Figure 13. IR emission spectra of systems $(Br^-)_i (H_2O)_{50-i} + i\,H_2O + 6\,X$: (a) $X = O_2$, (b) $X = O_3$; (1) $i = 0$, (2) $i = 2$, (3) $i = 4$, (4) $i = 6$; (5) $P(\omega)$ function of the $(H_2O)_{50}$ cluster.

Chapter 4

ESTIMATION OF THE GREENHOUSE AND ANTI-GREENHOUSE EFFECTS

An anti-greenhouse effect can be created due to the reduction of scattering centers at the formation of water clusters. The curve 1 in Figure 14a shows the relative change of integrated absorption intensity of IR radiation at the formation of $(H_2O)_{20}$ cluster by successive addition to dimer of Δn water molecules. The line 2 shows the relative total intensity of IR radiation absorption by $(H_2O)_{10}$ cluster and by five $(H_2O)_2$ dimers. The result of agglomeration of these clusters is again $(H_2O)_{20}$ cluster. In the case of water there is a double decrease of the greenhouse effect at the cluster formation (Figure 14a). The reduction of amount of absorbing centers gives the main contribution to the anti-greenhouse effect. Besides, the growth of water cluster is accompanied by the decrease of its absorbing ability (the curve 1 is almost everywhere in the area of negative values) due to the change of frequency characteristics of the total dipole moment. The addition of hydrocarbon molecules, as a rule, strengthens the greenhouse effect (Figure 14b). When one or two CH_4 molecules are added to $(H_2O)_{20}$ cluster the relative integrated intensity of IR radiation absorption weakens. The anti-greenhouse effect connected with the association of clusters surpasses the effect obtained from the change of vibration frequency of the dipole moment by ten times.

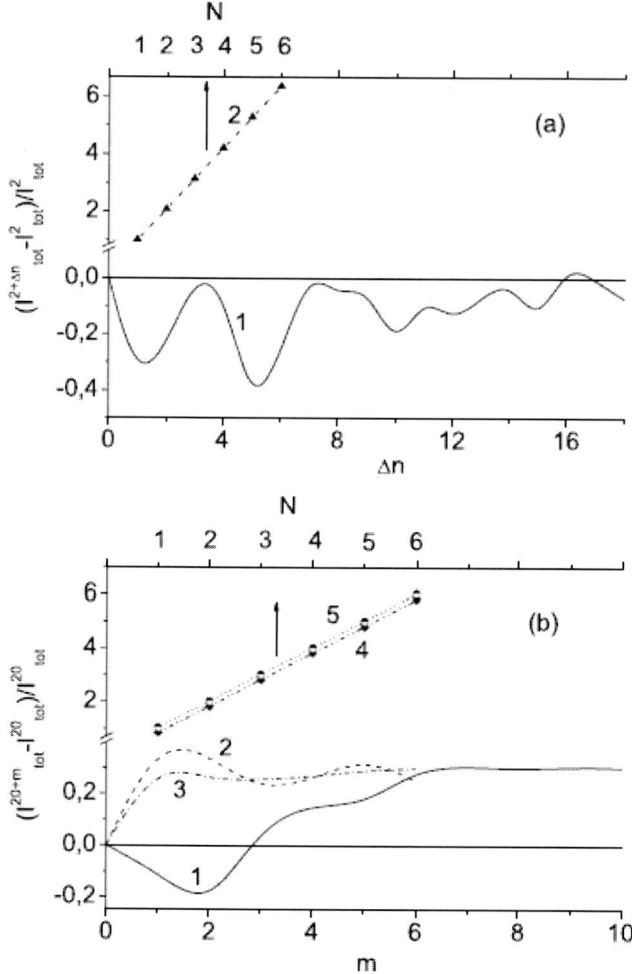

Figure 14. (a): *(1)* the relative integrated intensity of water dimmer IR absorption growing due to the addition of Δn H_2O molecules and *(2)* the total relative integrated absorption intensity of N clusters, $N_1 - (H_2O)_{10}$, $(N_2 - N_6) - (H_2O)_2$; (b): the relative integrated intensity of IR absorption by $(H_2O)_{20}$ cluster which joins m molecules: *(1)* CH_4, *(2)* C_2H_2, *(3)* C_2H_6 and the total relative integrated absorption intensity of N clusters: *(4)* $N_1 - (CH_4)_8(H_2O)_{10}$, $(N_2 - N_6) - (H_2O)_2$; *(5)* $N_1 - (CH_4)_3(H_2O)_{10}$, $(N_2 - N_6) - CH_4(H_2O)_2$.

The X-ray investigations show that there are dense structures – clusters in liquid water. If there were no such structures the density of liquid water would be 1.8 g/cm^3 instead of 1.0 g/cm^3. The exchange of molecules between the non structural water and clusters is constantly carried out so that the lifetime of clusters on average makes ~ 100 ps. The fraction of non bonded in clusters molecules in water is about 30 % at 20° C. The average size of cluster decreases with the increase of temperature and the fraction of non bonded molecules grows. Abnormal high heat capacity of water is explained by the melting of clusters. Atmospheric clusters are subject to less intensive external influence than clusters forming water. Therefore their average lifetime $\bar{\tau}$ should exceed 100 ps which can be considered as the bottom limit for $\bar{\tau}$. Because of the big number of uncertainties it is difficult to find the top limit for $\bar{\tau}$ in the atmosphere. However it should not exceed the period of full replace of water in the atmosphere, i.e. ~ 9 days. The consideration of the H_3O^+ ions in the atmosphere gives certain estimation of lifetimes [74]. Lifetime of the $H_3O^+(H_2O)_n$ particles is basically defined by concentration of free electrons (e^-). Reaction

$$H_3O^+(H_2O)_n + e^- \rightarrow H + (H_2O)_{n+1}$$

reduces lifetime of the $H_3O^+(H_2O)_n$ particles at heights more than 85 kms up to approximately 10 minutes. This time is quite enough for a considerable fraction of $H_3O^+(H_2O)_n$ particles to penetrate from heights more than 85 kms into an interval of 85-20 kms and participate in the process of turbulent diffusion by large whirlwinds. Here the concentration of electrons is less by several times and consequently the lifetime of $H_3O^+(H_2O)_n$ increases by several orders.

4.1. SPECTRAL PROPERTIES OF ATMOSPHERIC GASES

Atmospheric gases are selective to the wavelength of solar light which they absorb. Of all atmospheric gases, ozone and water vapor absorb the larger part of solar energy. Ozone absorbs the energy of ultraviolet radiation a small part of energy of visible light in the spectral range of 0.4–0.56 μm. Water vapor weakly absorbs radiation at wavelengths between 0.7 and 4 μm. Carbon dioxide very weakly absorbs solar energy, similarly to methane and chlorofluorocarbons. A large part of energy of visible radiation of the Sun is transferred to the surface, while the radiation at wave-length below 0.28 μm does not reach the surface. The amount of solar energy absorbed by the atmosphere largely depends on the amount of ozone and water vapor it contains.

Furthermore, atmospheric gases selectively absorb radiation from the Earth surface. Water vapor absorbs the energy of this radiation in the wavelength range of 5-8 μm and over 12 μm. Carbon dioxide and ozone radiate and absorb the energy at wavelengths in the vicinity of 15 μm and 9.6 μm, respectively. Atmospheric gases weakly radiate and absorb energy in the region of 10–12 μm. This region relates to atmospheric infrared window, because the atmosphere in this range actually does not absorb the IR radiation of the surface.

Clouds well reflect the solar energy. They also well reflect and absorb the long-wave energy in the wave-length range from 10 to 12 μm. Therefore, the window effectively closes in the presence of clouds. The Earth surface radiates waves of 10–12 μm, so that in the case of clear sky a large part of radiation goes to outer space. The frequently observed global warming or intensification of greenhouse effect is associated with the increase of carbon dioxide in the atmosphere. Carbon dioxide strongly absorbs radiation in the infrared spectral region; however, its content in the atmosphere is low and, therefore, its absorption is by and large insignificant. Atmospheric aerosols, i.e., clouds and solid particles suspended in the atmosphere, well absorb the solar radiation as well. By and large, 15–20% of radiation coming from the Sun to Earth is absorbed in the atmosphere. In addition to absorption, direct solar radiation on the way through the atmosphere is weakened by scattering, this latter weakening being more significant. Aerosol impurities scatter about 25% of energy of the total flux of solar radiation.

4.2. Energetics of the Atmosphere

The most important way of energy exchange between Earth and outer space is by radiation. The radiation balance is made up by the incoming energy of the Sun, the energy reflected back into outer space, and the energy radiated by Earth to outer space. The year average characteristic of radiation balance is usually considered. The total value of solar energy flux incident per square

Estimation of the Greenhouse and Anti-Greenhouse Effects 45

meter of area above the atmosphere is 342 W. The average annual albedo of the planet is 30%. Therefore, approximately 30% or 107 W/m^2 of incoming flux of solar energy goes back to outer space. The remaining 70% (~235 W/m^2) is absorbed by the atmosphere and surface. Each square meter of the atmosphere absorbs 168 W. The Earth surface acquires long-wave energy and radiates long-wave energy as well. On the average, the radiation power of each square meter of the Earth surface is 390 W. Some part of the energy radiated by the surface goes to outer space, and the other part is absorbed by the atmosphere. It is only 40 W/m^2 of radiation power of the Earth that passes through the atmosphere directly to outer space. The atmosphere absorbs the specific power of 350 W/m^2. The atmosphere radiates energy in all directions, including the radiation to outer space and towards the surface. On the average, each square meter of area of the atmosphere radiates the power of 195 W to outer space and 324 W toward the surface [7]. Therefore, the radiation from the atmosphere which reaches the Earth surface has a power which is 1.93 times that of the solar radiation directly heating the Earth surface. In addition, longer-wave radiation comes to Earth from the atmosphere and is readily absorbed by molecules of greenhouse gases and atmospheric clusters. The loss of long-wave energy of the surface exceeds the energy input, i.e., by and large the surface loses 66 = (324 − 350 − 40) W per square meter of area. The loss of atmospheric infrared power of 351 = (324 + 195 − 168) W/m^2 exceeds the power input (350 W/m^2) from the surface. The net contribution made by Earth radiation to outer space is 235 W per square meter of the surface; in so doing, the increment of energy of solar radiation and the infrared loss above the atmosphere are balanced. When the solar energy input prevails over the loss of long-wave energy, the surface has the net contribution to radiation energy of 102 = (168 − 350 − 40 + 324) W/m^2, and the atmospheric loss of radiation energy amounts to −102 = (67 + 350 − 324 − 195) W/m^2. The loss of 102 W per square meter of area of the atmosphere is equivalent to the cooling of atmosphere by more than 200 K during one year. No such strong cooling is observed, because the energy is transferred from the surface to atmosphere in other than radiation ways as well. The transfer of 102 W/m^2 of power is associated with two forms of heat transfer, namely, tangible and latent ones. The tangible heat transfer is a combination of processes of heat conduction and convection and is of the order of 24 W/m^2. The latent heat transfer from the surface to atmosphere is about 78 W/m^2. The evaporation from the oceans and water basins and the sublimation from the glaciers cool down the surface. A part of water evaporating to the atmosphere condenses and forms clouds and rainfall while liberating latent heat. The balance between the energy input and loss is disturbed when its annual averaging is performed depending on latitude. The tropical regions give a radiation energy input, while the polar regions are characterized by cooling associated with radiation. Because the tropics are not overheated and the polar regions are not overcooled, the energy must be transferred from the tropics to poles. This transfer is realized by atmospheric and oceanic flows which largely define the weather.

If the energy balance of the planet changes as a result of human activities, the atmospheric and oceanic heat transfer will have to be varied for recovering this balance. In particular, the energy balance of the planet may change as a result of variation of composition of the atmosphere. Apparently, the feedback is possible. For example, assume that all atmospheric water clusters are at $T = 273$ K. This corresponds on the average to the air temperature at an altitude of about 3 km. Then the power of 130 W/m^2 will be required for evaporating all water clusters in the atmosphere during one day. This power amounts to ~40% of power (324 W/m^2) radiated by the atmosphere to the Earth surface and is 27% higher than the power (102 W/m^2) defining the balance between radiation of the atmosphere and of the Earth surface. Therefore, an instantaneous disturbance of thermal balance may significantly affect the content of clusters in the atmosphere even within several days.

4.3. Effect of Space Engineering

The environmental importance of the effect made by space engineering on the upper atmosphere is evident. The frequency and intensity of glows in the upper atmosphere considerably increased recently. Mainly lithium, sodium, and aluminum oxides fluoresce at an altitude of about 100 km, and OH^- ions – at 250 km. Sodium, trimethyl aluminum, aluminum oxide, chlorine, hydrogen chloride, nitrogen oxide, carbon dioxide, water, hydrogen, strontium, cesium, barium, and other substances are released to the atmosphere during rocket flights. Up to 520 rocket launchings a year are made in the world. Discharges from one launching of Shuttle amount to 1261 t of reagents, and those of Energiya – to 1490 t. The discharges from Energiya contain carbon oxides, water, and hydrogen; in addition to these components, the discharges from Shuttle include chlorine, hydrogen chloride, and aluminum and nitrogen oxides. During the period of 1982–1990 alone, the launchings of Shuttle resulted in discharges to the upper atmosphere of 6358 t of chlorine and hydrogen chloride, 238 t of nitrogen oxide, 12 852 t of carbon oxides, 8704 t of water and hydrogen, and 6018 t of aluminum oxides. The growth of rocket-and-space engineering leads to emergence of large-scale atmospheric processes, in particular, to depletion of ozone layer of the stratosphere (Figure 15). Further degradation of ozone layer will cause a temperature decrease in the stratosphere and heating of close-to-Earth layer of the troposphere [75].

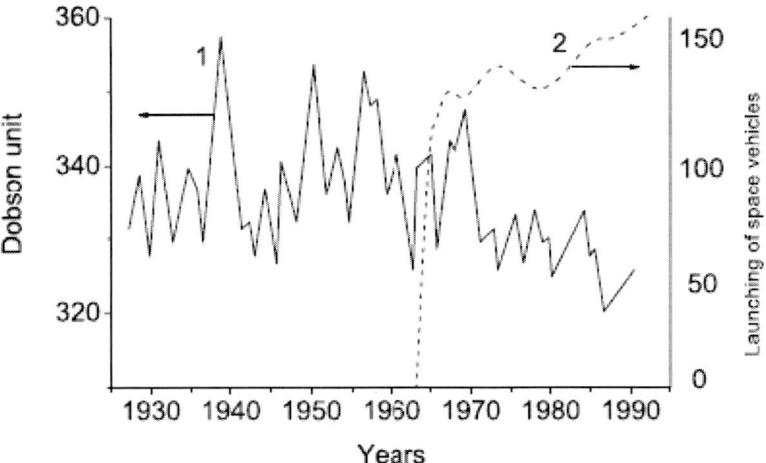

Figure 15. The correlation between the loss of ozone in the atmosphere and the number of space vehicles launched from the Earth [75]: (*1*) reduction of ozone content in the stratosphere, (*2*) dynamics of launching of space vehicles.

4.4. ABSORPTION OF INFRARED RADIATION BY CLUSTERS

Because the water vapor in the atmosphere is quite rarefied, we can assume that the free energy of such vapor is zero. Then, for determining the number of water clusters of each type, we will assume that the probability of revealing the clusters is proportional to the Boltzmann factor $f_B = \exp(F/kT)$. The values of free energy of water clusters containing from two to eight molecules were borrowed from Shenter et al. [76], and the analogous characteristic of clusters having from 9 to 20 molecules was calculated by the data of MD calculation of Gonzales et al. [77]. The values of Boltzmann factor for clusters sizes in the range of 2–20 molecules are given in Figure 16, and the values of corresponding free energy per molecule are given in the inset of this figure. One can see that the specific free energy decreases with increasing cluster size. This causes a rapid decrease in the Boltzmann factors of growing clusters and, therefore, of the probabilities of their detection. The ratio of Boltzmann factors for clusters of 20 and 2 molecules is 4×10^{-4}. Therefore, in the case of thermodynamic equilibrium, clusters of more than 20 molecules in size may be ignored for estimation purposes. The dependence $f_B(n)$ obtained by Mhin et al. [78] decreases even more rapidly with increasing n because of using lower values of free energy.

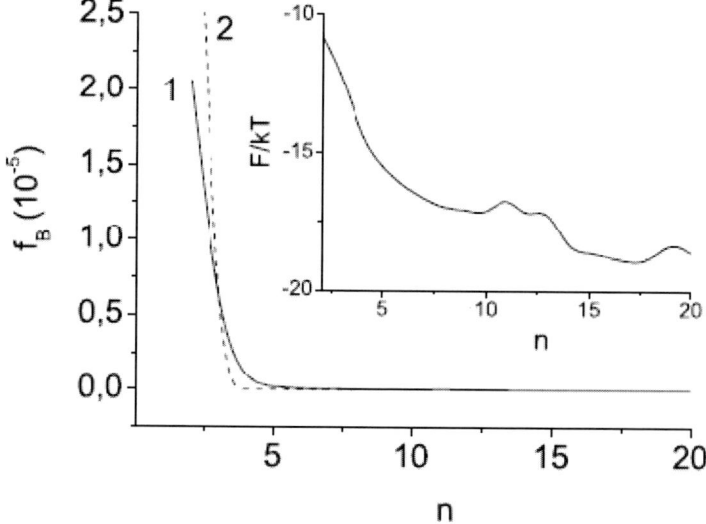

Figure 16. The Boltzmann factors for water clusters containing n molecules: (*1*) our calculation, (*2*) data of [78]. Given in the inset is the relative free energy of water clusters per molecule.

The data on the absorption of IR radiation by water clusters were obtained by the method of MD simulation in [40-42, 79, 80]. The clusters considered here exhibit a lower integral intensity of absorption of IR radiation than one water molecule. The relative (with respect to water molecule) absorptivity of water clusters is demonstrated by curve *1* in Figure 17a. An analogous characteristic with respect to water dimer is given by curve *2*. A water dimer absorbs IR radiation stronger than H_2O molecule; as a result, curve *2* in Figure 17a is located below curve *1*. One can see that the quantity I_{cl}/I_1 (where I is the integral intensity of absorption of IR radiation, subscript *cl* indicates water clusters, and subscript 1– water molecule) hardly increases with the cluster size, and all relevant values are $I_{cl}/I_1 < 1$. Specific intensity of absorption i_σ of IR radiation (i.e. the intensity of absorption per 1 molecule in cluster) decreases with growth of the size of the water aggregate (Figure 17b). The character of change of function $i_\sigma(n)$ is not monotonous. The approximation of dependence $i_\sigma(n)$ by a polynom of 3rd degree is shown on Figure 17b by a dotted line. The greatest deviation of values i_σ from appropriate polynom dependence is observed for clusters containing 5, 7 and 9 water molecules. Clusters formed by 5 molecules of water have significantly higher susceptibility to IR radiation in comparison to the one determined by the smoothed dependence (dotted line). Efficiency IR radiation absorption by water clusters falls with the growth of cluster size. Specific absorption in relation to individual H_2O molecule weakens 21.5 times for a cluster consisting of 20 water molecules. It is a general tendency for phenomena determined by behaviour of a collective variable, in this case by the total dipole moment of a cluster. It is of interest to estimate efficiency of IR radiation absorption by water molecules which is taking place in the Earth's atmosphere and on its surface. It is known that the bulk of water in the Earth's atmosphere makes about 1.27×10^{16} kg, and its quantity on the Earth ~ 1.35×10^{21} kg. Taking into account that the weight of water molecule equals to 2.99×10^{-26} kg, we find approximate number of water molecules in the atmosphere (0.424×10^{42}) and on the Earth (0.451×10^{47}). Total specific power of radiation absorbed by the atmosphere makes 417 W/m² (67 [from Sun] and 350 [from Earth]). Specific absorption of radiation by the Earth equals 492 W/m² (168 [from Sun] and 324 [from atmosphere]). The area of the Earth's surface is defined by the size (510099699.07 km²) and the fictitious shell in the stratopause is characterized by the area 518137426.17 km². The radius of a fictitious shell is chosen equal to half of the distance from the top border of thermosphere to the surface of the Earth, i.e. 50 km. Water on the Earth occupies territory 2.5 times larger than land. Thus, the area of water surface on the Earth is calculated as 2.5/3.5 = 0.71 aside from all surface of the planet or 364356927.90 km². The total power accumulated by atmospheric moisture is 216.06 GW, and water on the Earth absorbs 179.26 GW. As a result $508924.93 \times 10^{-36}$ W of absorbed radiation power is required for one water molecule in the atmosphere, and only 3.97×10^{-36} W for one H_2O molecule in the water pool of the Earth. Thus, water molecule in the atmosphere, on average, absorbs 128120 times more acting radiation, than H_2O molecule in oceans or rivers. Further we shall show, that more than two thirds of water molecules in the atmosphere are monomers and the rest form a disperse media consisting of clusters, drops or microcrystals. At the same time all H_2O molecules of the water pool are connected amongst themselves and represent huge weights of water or ice.

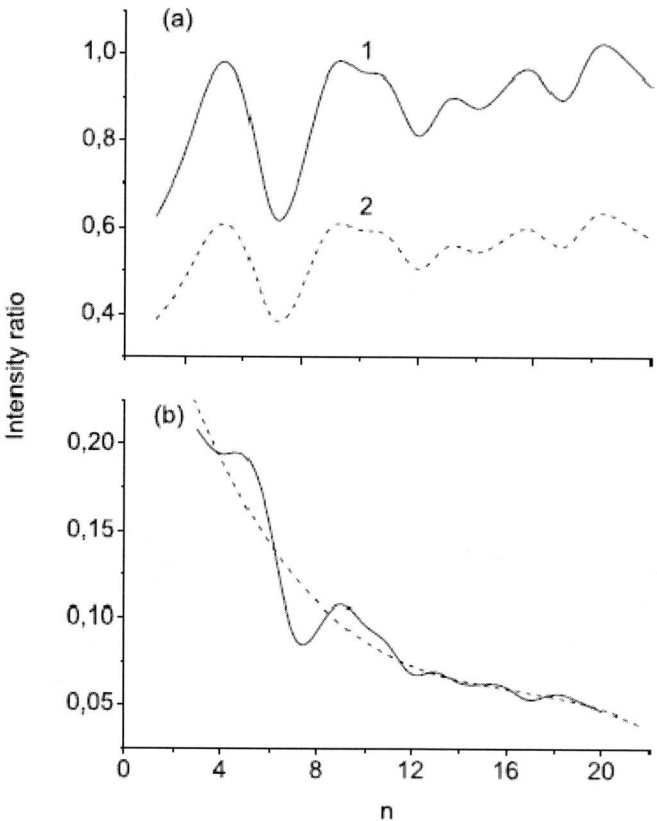

Figure 17. (a): The ratio of integral intensity of absorption of IR radiation by water clusters formed by n molecules to the respective characteristic of (*1*) free molecule of water and (*2*) water dimer at average moisture content of 11 g/m^3; (b): the intensity of IR radiation absorption per 1 molecule i_σ in the cluster (solid line) and approximation of dependence $i_\sigma(n)$ by the polynom of 3rd degree (dotted line), n is number of water molecules in cluster.

After determining the total mass of clusters M_{cl} and their statistical weights, the number of clusters $N_{cl}^{(i)}$ of each type was determined,

$$N_{cl}^{(n)} = \frac{M_{cl} W_n}{m_n},$$

where the statistical weight of cluster was determined as $W_n = \dfrac{f_B^{(n)}}{\sum_{n=1}^{20} f_B^{(n)}}$, and m_n is the mass of cluster containing n molecules of water. The number of all clusters was given by the sum $N_{cl}^{tot} = \sum_{i=1}^{20} N_{cl}^{(n)}$.

The overall absorptivity of clusters with respect to integral intensity of absorption I_1 of one water molecule was calculated as

$$\alpha^* = \sum_{n=2}^{20} N_{cl}^{(n)} \alpha_n^*,$$

where $\alpha_n^* = I_{cl}^{(n)} / I_1$ is the relative absorptivity of cluster consisting of n molecules of water. In what followed, the greenhouse effect of water vapor was estimated by comparing α^* with the number of water monomers because, in the case of free water molecules, $\alpha^* = 1$.

The relative combined intensity of absorption of IR radiation $\sum_n I_{cl}^{(n)} / I_1$ as a function of cluster size is given in Figure 18 for three values of average surface humidity ρ_0^w. The main contribution to absorption is made by clusters of the first three-four types in accordance with their numbers. At present, ρ_0^w amounts to approximately 11 g/m^3. The absorption of IR radiation by clusters will be reduced by a factor of 1.75 as a result of variation of humidity ρ_0^w from 11 to 15 g/m^3. This is associated with redistribution absorption of IR radiation between moisture-containing components with increasing ρ_0^w. The increase in ρ_0^w from 11 to 15 g/m^3 will cause an increase in the total amount of moisture and vapor of monomers in the atmosphere by a factor of 1.36, the mass of ice will increase by a factor of 1.56, that of droplets – by a factor of 2.96, and that of clusters – by a factor of only 1.22. As a result, the clusters will make a significantly smaller contribution to absorption of IR radiation. It is known that clouds cover about one half of the Earth surface. The average cloud thickness may be taken to be 1 km. We will assign to the liquid phase the droplets whose size exceeds 0.5 μm, i.e., the minimal size of droplet observed in clouds. The water content of droplet liquid clouds may fluctuate from 0.2 to 8 g/m^3. For estimation purposes, we take 4 g/m^3 as the averaged value of water content; then, the estimated value of mass of droplets and crystals in the atmosphere is 1020 Gt. This value adequately agrees with the mass of condensed phase in the atmosphere determined below (1530 ± 230 Gt). The average surface humidity ρ_0^w is related to water content x of effective clouds covering half the Earth by an almost linear increasing dependence (Figure 19). Here, the contribution to effective cloud 1 km thick is made by both droplets and crystals.

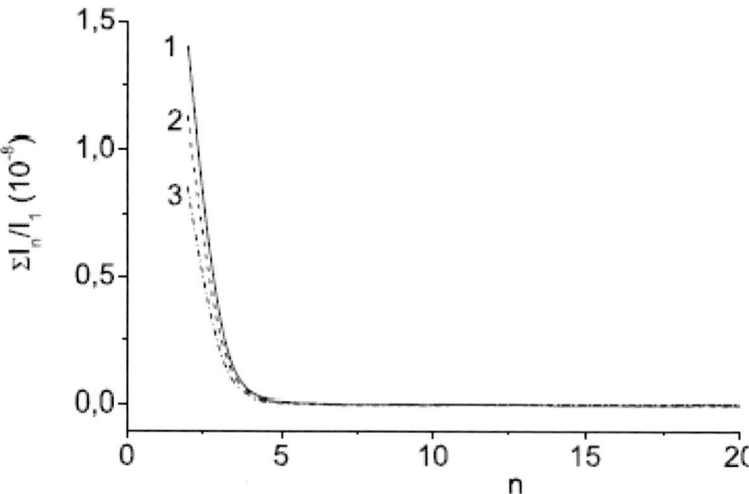

Figure 18. The combined integral intensity of absorption of IR radiation by water clusters containing n molecules in absorption units of free water molecule at average surface humidity of (1) 11 g/m^3, (2) 13, and (3) 15.

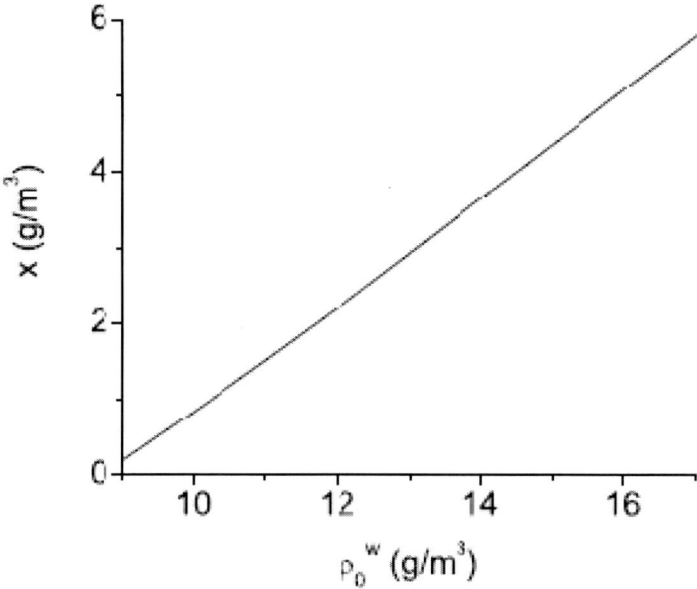

Figure 19. Correlation between water content x of effective clouds and average surface humidity ρ_0^w.

4.5. EMISSION SPECTRA OF CLUSTERS

As distinct from absorption spectra, emission spectra are more frequently studied under nonequilibrium conditions using various methods of excitation [81]. In the simplest case of optical excitation, an atom from the ground state goes to a higher discrete energy level. The excitation of water molecules in a cluster is possible as a result of absorption of molecules of other gases by the cluster. The Earth atmosphere is largely made up by nitrogen (78.08%), oxygen (20.94%), and argon (0.93%). The fraction of the remaining gases is 0.05%, of which 0.03% falls on CO_2. The emission spectra $P(\omega)$ of ultradisperse water systems formed by clusters of $(H_2O)_n$, $10 \leq n \leq 50$, and by molecules of main atmospheric gases (N_2, O_2, and Ar) were calculated by the molecular dynamic method using the method of flexible molecules [82–84]. The inclusion of intramolecular vibrations made it possible to expand the frequency range of spectra $P(\omega)$. Figure 20a gives spectra for disperse water systems, with each cluster of these systems absorbing one gas molecule, and Figure 20b – spectra for systems with two absorbed gas molecules in each cluster. The absorption of molecules of atmospheric gases leads to a strong excitation of water clusters, especially when these clusters capture Ar atoms. The single absorption of N_2 and O_2 molecules and Ar atoms by water clusters defines the following ratio between integral intensities I_P of $P(\omega)$ spectra: 1:7.6:36.1. In the case of double absorption of N_2 molecules, the intensity of emission spectrum increases; when each cluster absorbs two O_2 molecules or two Ar atoms, this intensity decreases. In this case, the ratio between the values of I_P is as follows: 1:2.2:8.5. After each water cluster absorbed one N_2 molecule, the integral intensity of spectrum $P(\omega)$ increased by a factor of 4.4. Some people believe that rather large ($n \geq 10$) water clusters in the atmosphere at altitudes from 10 to 30 km (i.e., at $T < 230$ K) may exist for as long as possible in view of the fact that the energy of evaporation of molecule makes up the bulk of energy of the entire cluster [74]. In this case, the greenhouse effect generated by clusters is defined both by the intensity of absorption of IR radiation and by the rate of scattering of energy input or by the power of emission of thermal energy. In a thermally nonequilibrium gas, where no equilibrium exists between the translational and vibrational degrees of freedom of molecules, the time of radiation of $CO_2 - N_2 - Ar$ mixture (0.096:0.003:0.901) does not exceed 50 µs [85]. Apparently, the time of radiation of the same order of magnitude is to be expected for ultradisperse water systems which absorbed N_2 and O_2 molecules or Ar atoms. Therefore, the absorption of molecules of main atmospheric gases by clusters containing ten or more water molecules at $T < 230$ K must bring about amplification of the greenhouse effect generated by these clusters. However, such clusters may be injected into the above identified zones of atmosphere largely owing to combustion of rocket fuel. At present, the mass of such clusters does not exceed 0.005% ($\sim \{10^5 / [2.30 \times 10^9]\} \times 100\%$) of the mass of all clusters existing in the Earth atmosphere, and the greenhouse effect generated by the "rocket" clusters must be less than 5.5×10^{-5} K.

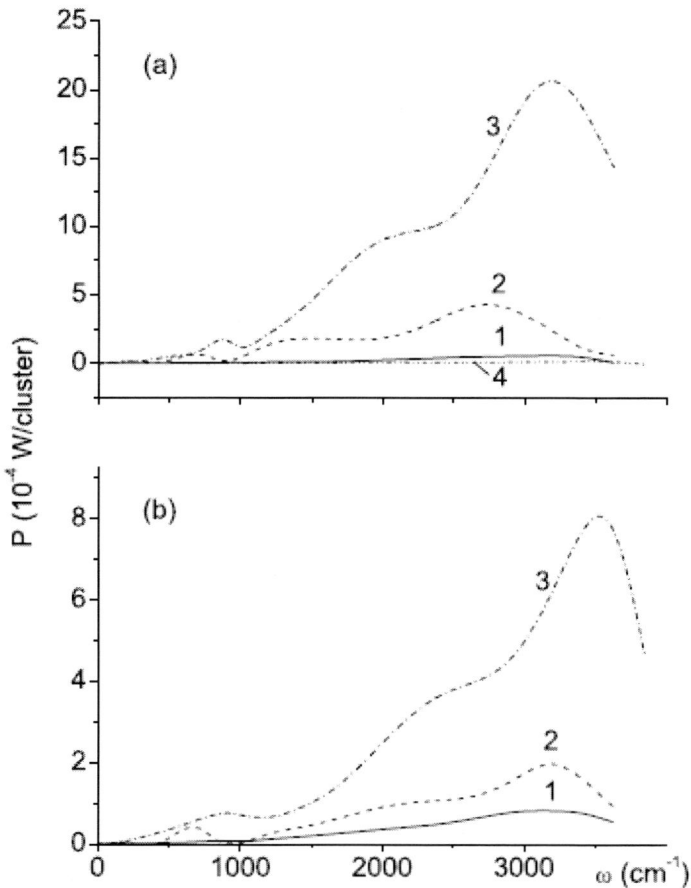

Figure 20. Emission spectra of cluster systems. (a): (*1*) $N_2(H_2O)_n$, (*2*) $O_2(H_2O)_n$, (*3*) $Ar(H_2O)_n$, (*4*) $(H_2O)_n$, $10 \leq n \leq 50$; (b): (*1*) $(N_2)_2(H_2O)_n$, (*2*) $(O_2)_2(H_2O)_n$, (*3*) $(Ar)_2(H_2O)_n$.

4.6. ANTI-GREENHOUSE EFFECT

Atmospheric water clusters absorb the IR radiation of the Earth and thus produce the greenhouse effect, the way it is done by free (not clustered) molecules of greenhouse gas. However, the intensity of radiation absorbed by clusters is not at all proportional to the number of constituent molecules [86].

Moreover, the clusters absorb the energy of incoming radiation in an amount comparable with, and sometimes even smaller than, the value of absorbed radiation energy given by free water molecule. As a result, the molecules making up the cluster when in free state absorb the IR radiation (of total intensity $\sum_i I_i$) much stronger than the entire cluster does (I_{cl}). The difference between $\sum_i I_i$ and I_{cl} is defined as the anti-greenhouse effect produced by cluster. This quantity may be expressed in degrees. The presently existing anti-greenhouse effect generated by atmospheric water clusters may be estimated as follows. The distribution of atmospheric humidity in altitude obeys the empirical dependence of Ghan [87],

$$\rho = \rho_0 \times 10^{-h/6.3}, \qquad (8)$$

where ρ_0 is a constant corresponding to humidity at certain altitude (as a rule, immediately at the Earth surface), and h is the altitude above the Earth surface in km. On the one hand, formula (8) may be used for calculating the distribution of vapor of water monomers above the Earth surface, and on the other hand – for determining the altitude distribution of the overall amount of moisture in the atmosphere. In the former case, the constant ρ_0 is provided by the fraction of water monomers in the vicinity of the Earth surface, which was found by extrapolating the Boltzmann size distribution of clusters towards the value of $i = 0$. The thus determined value of ρ_0 amounted to ~66.3% of the known (~11 g/m^3) value of humidity, which is defined at the surface by the number of water monomers and clusters. In the latter case, the value of ρ_0 corresponded to humidity at an altitude of 3 km, determined as a result of spectroscopic measurements [88]. Experimental measurements were performed in the presence of clouds at an altitude of 1 to 2 km. The spectroscopic sensing element enables one to measure spatially separated profiles of moisture density both around tropospheric clouds and within these clouds. The integration of the first and second distributions over concentric layers of thickness h gives the values of mass M_{vap} of monomer vapor and of total amount M_{tot} of moisture in the atmosphere. The mass of droplets and crystals in the atmosphere was found using the formula

$$M_{\text{drop(cryst)}} = \sum_{n=1}^{100} \left(\rho^{(n)} - \rho_{sv}^{(n)} \right) V_n,$$

where $\rho^{(n)}$ and $\rho_{sv}^{(n)}$ denote the density of moisture and of saturated water vapor in the nth layer 1 km thick, and V_n is the volume of the nth layer. Water droplets and ice crystals are formed in clouds when $T < 273$ K; however, in the former case it is necessary that air would be supersaturated with respect to water, and in the latter case – with respect to ice. Because the saturation curves of water vapor above the surfaces of water and ice are quite close, both droplets and ice crystals may be contained in one and the same cloud. Apparently, the total mass of ice in clouds is proportional to the concentration c'_{crys} of crystal nuclei in clouds because it represents the amount of crystal ice on the condensed phase of clouds. Linear laws of variation of $c_{\text{crys}} = c'_{\text{crys}} /(c'_{\text{crys}} + c'_{\text{liq}})$ with decreasing temperature beginning at 273 K were suggested in [89, 90]. The mass of ice crystals was determined in accordance with both dependences $c_{\text{crys}}(T)$ given in Figure 21. The difference between the masses of ice determined using dependences 1 and 2 in Figure 21 amounted to 30%. The use of dependence 2 leads to a higher value of M_{crys} (1005.27 Gt).

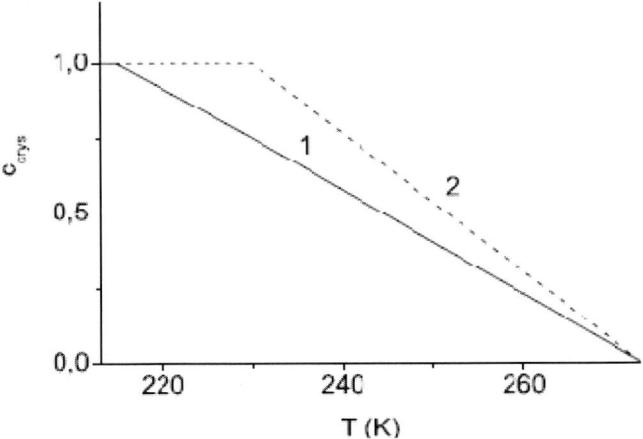

Figure 21. The fraction of ice crystals in the condensed phase of clouds, determined in accordance with temperature dependences of ice concentration in (1) [89] and (2) [90].

In determining the mass of droplets in the atmosphere, $\rho_{sv}^{(n)}$ was provided by $\rho_{svl}^{(n)}$, i.e., the density of saturated vapor over supercooled water, and the quantity $\rho_{svc}^{(n)}$, i.e., the density of saturated vapor over ice at the temperature of the nth layer, was used for calculating the mass of crystals. The temperature dependences of $\rho_{svl}^{(n)}$ and $\rho_{svc}^{(n)}$ are given in Figure 22 as curves *1* and *2*, respectively. Curve *1* was constructed using the temperature dependence of density of saturated vapor over supercooled water ρ_{svl}, which was assigned as

$$\rho_{svl} = 4.93 \times 10^{8.61503(T-273.15)/T},$$

and curve *2* —using the dependence $\rho_{svc}(T)$ of density of saturated vapor over ice, which was given as

$$\rho_{svc} = 4.93 \times 10^{9.76421(T-270.15)/T}.$$

Here, the values of ρ_{svl} and ρ_{svc} are in g/m³. The $\rho_{svl}(T)$ and $\rho_{svc}(T)$ curves rapidly decrease (by a factor of ~30–40) with the temperature varying from 283 to ~240 K and then gradually decrease with further temperature drop. Because the saturated vapor pressure of liquid water is always higher than the saturated vapor pressure of ice, the ice crystals will grow due to consumption of liquid water. For example, air saturated with respect to liquid water becomes supersaturated with respect to ice by 10% at 263 K and by 21% at 253 K. This will result in a rapid transition of liquid water to ice. As a result, the masses M_{drop} and M_{cryst} were found. The mass of clusters in the atmosphere was determined as

$$M_{cl} = M_{tot} - M_{vap} - M_{drop} - M_{cryst}.$$

Similarly, the altitude profile of density of clusters was determined in terms of respective distributions. The calculated $h(M)$ profiles demonstrate that the bulk of water clusters in the atmosphere are located up to the altitude of 2 km (Figure 23). The planetary boundary layer (PBL) – the lowest 2 km or so of the atmosphere – is in frequent contact with the surface. The relative humidity in this layer is widely expected to remain fairly stable under climate changes, at least over oceans, because significant changes in its humidity would drive unsustainable changes in surface evaporation [91]. Overwhelming quantity of drops is concentrated within the limits of the 2.5 km altitude, and crystals are formed, from the altitude of 3 km. Above the PBL it has been less clear what controls humidity [92]. The air there is not always in frequent contact with the surface, its specific humidity is highly nonuniform, and relative humidity often reaches values near 1% or less. Water vapor Raman lidars detect the light backscattered by nitrogen and water vapor molecules. The ratio of the photons scattered by water vapor and nitrogen is proportional to the water vapor mass mixing ratio. The humidity profiles, observed by Koldewey Aerosol Raman Lidar (KARL), show a nearly homogenous humid layer up to 2 km altitude [93]. The mixing ratio amounts up to about 1.5 g/kg. Above 2.3 km the humidity decrease strongly, reaching less than 0.1 g/kg at 4 km altitude. Probably, local effects created by water clusters and observed by stationary lidar mostly cannot be resolved by satellite soundings or atmospheric models used for meteorological analyses or regional climate investigations.

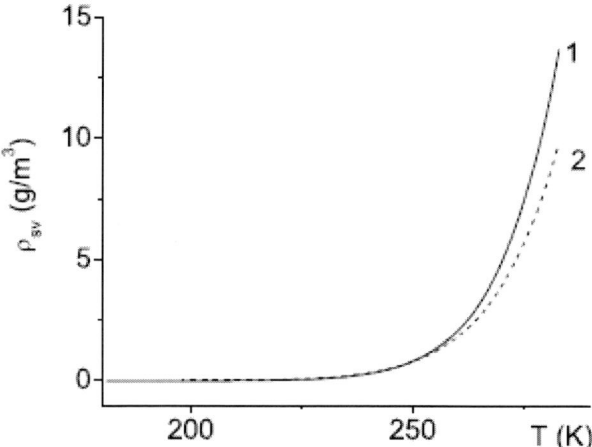

Figure 22. The saturated vapor density as a function of temperature in the temperature range of the Earth atmosphere: (*1*) vapor over supercooled water, (*2*) vapor over ice.

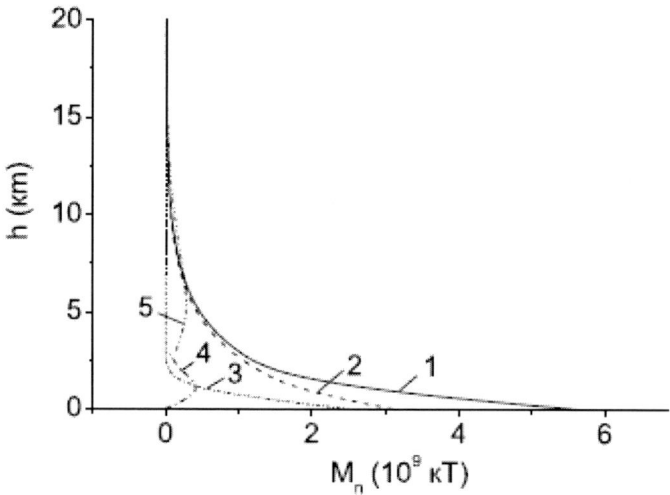

Figure 23. Contributions to a total mass of the atmosphere moisture: *(1)* all moisture, *(2)* monomers vapor, *(3)* clusters, *(4)* droplets, *(5)* crystals.

If the overall amount of moisture in the atmosphere is taken to be 100%, the ratio between the amounts of monomer vapor, clusters, droplets, and crystals appears to be 70.0:18.0:5.9:6.1 in case the mass of crystals is found according to [89] and 70.0:16.2:5.9:7.9 in case the mass of crystals is found according to [90] (Figure 24). On the average, the integral intensities of absorption of IR radiation by water clusters are lower than those by a single molecule. The estimated number of clusters in the atmosphere is 2.4 times smaller than the number of molecules which form these clusters. The anti-greenhouse effect of clusters is defined as the difference between the increments of the average global temperature of the Earth surface caused by the absorption of IR radiation by free water molecules which make up the clusters and the absorption of the clusters proper. The greenhouse and anti-greenhouse effects produced by clusters amount to ~1.1 ± 0.1 and 3.3–3.6 K, respectively. The error of estimation of these effects is found as the difference between the respective characteristics assigned by the methods of determining the amount of the crystal phase. Therefore, the average temperature of the planet could rise by 3.6 K in the absence of clusters, which would result in a significant change of climate. The increase in the global average temperature of the Earth surface during the last 100 years did not exceed 0.74 K [94].

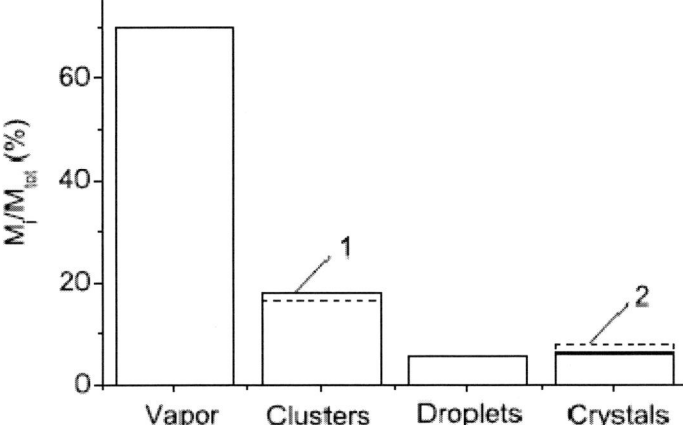

Figure 24. The ratio between monomer vapor, clusters, droplets, and crystals, obtained using the linear temperature dependence for the number of crystals in accordance with (*1*) [89] and (*2*) [90].

Chapter 5

Concluding Remarks

The absorption and scattering of infrared radiation during transmission through the Earth atmosphere lead to its attenuation. Atmospheric nitrogen and oxygen do not absorb IR radiation and only attenuate this radiation as a result of scattering, largely for visible light. Atmospheric moisture, carbon dioxide, ozone, and other impurities, which are present in the atmosphere, selectively absorb IR radiation. The infrared radiation is absorbed especially strongly by the water components of the atmosphere, whose absorption bands are located in almost the entire infrared spectrum. Carbon dioxide absorbs appreciably in the middle IR region. However, "windows" transparent to infrared radiation are still available in close-to-Earth layers of the atmosphere in the middle infrared region. The thermal balance may be disturbed both in the case of variation of concentration of some greenhouse component in the troposphere and in the case of serious disturbances of the existing structure within the entire atmospheric depth including the thermosphere. The Earth climate is a manifestation of absorption of radiation coming from the Sun and of reradiation to outer space. The resultant energy is redistributed by the atmosphere and the ocean. Any change in the balance of energy received by the Earth surface and radiated by this surface will immediately affect the climate. The composition of the atmosphere is one of the parameters influencing the energy balance of the Earth.

Absorption by disperse water system of IR radiation was slightly amplified as a result of N_2O molecules absorption and was weakened after absorption of CO_2 molecules. In the case of methane molecules' absorption the transparency window for IR radiation is observed. As a rule the absorption of GHG causes the increase of reflection coefficient of monochromatic infrared radiation, but coefficient R decreases in the case of the absorption of methane. Pure water clusters dissipate the energy most slowly. The radiation power of energy is increased after the addition to water aggregates of GHG molecules.

It is shown in the present work, that water clusters are capable to adsorb ozone and to detain its molecules on a surface during significant time. Clusters consisting of polar molecules can keep Cl^- and Br^- ions. Thus on a surface of water cluster favorable conditions are created for proceeding of reaction of chlorine or bromine with ozone which results in the destruction of the last. Duration of Br^- ions presence in water cluster is at least by 10 times more than this value for Cl^- ions. Therefore bromine ions destroy ozone more effectively. Intensity of IR absorption spectra essentially increases with the growth of Br^- ions' number per cluster. Raman spectra of disperse aqueous-ozone systems considerably reduce the intensity at increase of Br^- ions contained in them. Emission spectra of IR radiation essentially increase intensity only at significant concentration of Br^- ions in clusters.

Until recently there is no complete picture of all the factors influencing climate change. The reliability of forecasts depends on an understanding of cycles that result in climate change. For example, space rays represent the main source of ionization of the air. Under the influence of space rays nitrogen oxides are formed, and they destroy the ozone layer of the stratosphere which results in a substantial cooling. The total mass of ozone makes $\sim 0.64 \times 10^{-5}$ of the weight of all atmosphere. The depletion of the ozone layer can be connected to other natural phenomena, for example, volcanic eruptions. We can distinguish chloral, nitric and hydrogen cycles of the ozone destruction. Chlorine destroying ozone is not spent at all in this process. At destruction of N_2O molecules N_2 and O_2 are formed. The basic stocks of hydrogen are concentrated in the nucleus of the planet, and hydrogen through deep fractures emits into the atmosphere. Almost all ozone holes are placed above the seismic zones of the Earth.

It is not precisely known at present why climatic warming occurs. Therefore, it is necessary to study the processes occurring in the atmosphere more thoroughly. The present study deals with an atmospheric phenomenon such as anti-greenhouse effect resulting from the clustering of water vapor. A process of clusterization is accompanied by a sharp reduction of number of scattering centers. The greenhouse effect produced by water clusters is well known, but nothing was known about the anti-greenhouse effect. It was found that water clusters exhibit a lower integral intensity of absorption of IR radiation than the respective characteristic for one water molecule. If a cluster contains n molecules, it may absorb the energy of IR radiation which is at least n times lower than the energy absorbed by n free molecules. The anti-greenhouse effect is defined as the difference between the greenhouse effects of free molecules and of clusters formed by these molecules.

The total cooling effect of water evaporation from the surface of seas and reservoirs, is 13.4 K. Transition of the part of atmospheric water vapor into clusters causes a decrease of atmospheric temperature by 3.3–3.6 K, that is the effect of 25 ± 1 % cooling of the heat received during water evaporation. Thus, the content of CO_2 in the atmosphere of the Earth is not an absolute criterion of the efficiency of the greenhouse effect made by this gas which now is estimated at 9.3 K, i.e. constitutes about 19.6% of the effect given by all greenhouse gases. In the absence of reliable thermal protection from the ozone layer of the lower stratosphere and depletion of atomic oxygen in the thermosphere, CO_2 can partially or completely carry out the functions of anti-greenhouse gas.

The bulk of clusters are concentrated up to the altitude of 2 km, and the droplets are largely located at altitudes up to 2.5 km. The crystals are, as a rule, located above 3 km. Water vapor is most widespread among other components of atmospheric moisture. Water clusters are injected into the upper atmosphere during flights of space vehicles. Such clusters may exist for a long time; however, their fraction is small at present, and the greenhouse effect they produce is low.

As a whole, the warm climate period of the Earth is equivalent to approximately 10 % of the glacial cycle duration which in itself consists of approximately 100 000 years. Hence we are living in a warm climate period that has effectively lasted for 10,000 years. It is expected that in the near future cooling of the planet will take place.

REFERENCES

[1] Milankovich M. *Matematicheskaya klimatologiya i astronomicheskaya teoriya klimata* (Mathematical Climatology and Astronomical Theory of Climate); GONTI NKTP SSSR: Moscow, 1939.
[2] Barrett, J. *Energy & Environment* 2005, 16, 1037–1045.
[3] Held, I. M., and Soden, B. J. *Ann. Rev. Energy Environ.* 2000, 25, 441–475.
[4] Johnson, K. (2009). Water Buckyball Terahertz Vibrations in Physics, Chemistry, Biology, and Cosmology. [arXiv] e-print 2009. arXiv:0902.2035.
[5] Carlon, H. R., and Harden, C. S. *Appl. Opt.* 1980, 19, 1776–1786.
[6] Xie, F., Tian W., and Chipperfield, M. P. *J. Geophys. Res.* 2008, 113, D00B09, doi: 10.1029/2008 JD009829.
[7] *IPCC (Intergovernmental Panel on Climate Change) Changes in Atmospheric Constituents and in Radiative Forcing in Climate Change 2007*; Cambridge University Press: New York, 2007.
[8] Carlon, H. R. *Infrared Phys.* 1979, 19, 549–557.
[9] Gebbie, H. A. In Book *Atmospheric water vapor*; Deepak, A., Wilkerson, T. D. and Ruhnke, L. H.; Ed.; Academic Press: New York, 1980; pp 133–141.
[10] Low, G. R., and Kjaergaard, H. G. *J. Chem. Phys.* 1999, 110, 9104–9115.
[11] Goss, L. M., Sharpe, S. W., Blake, T. A., Vaida, V., and Brault, J. W. *J. Phys. Chem. A* 1999, 103, 8620–8624.
[12] Slanina, Z., and Crifo, J. F. *Int. J. Thermophys.* 1992, 13, 465–476.
[13] Goldman, N., Fellers, R. S., Leforestier, C., and Saykally, R. J. *J. Phys. Chem. A* 2001, 105, 515–519.

[14] Kulmala, M., Pirjola, L., and Makela, J. M. *Nature* 2000, 404, 66–69.
[15] Kulmala, M., Lehtinen, K. E. J., Laakso, L., Mordas, G., and Hameri, K. *Boreal Env. Res.* 2005, 10, 79–87.
[16] Hanson, D. R., and Eisele, F. L. *J. Geophys. Res.* 2002, 107, doi:10.1029/2001/JD001100.
[17] Akhmatskaya, E. V., Apps, C. J., Hillier, I. H., Masters, A. J., Watt, N.E., and Whitehead, J. C. *Chem. Commun.* 1997, 107, 707.
[18] Dunn, M. E., Pokon, E.K., and Shields, G. C. *J. Am. Chem. Soc.* 2004, 126, 2647–2653.
[19] Dang, L. X., and Chang, T-M. *J. Chem. Phys.* 1997, 106, 8149–8159.
[20] Galashev, A. Y., and Rakhmanova, O. R. *J. Struct. Chem.* 2005, 46, 626–632.
[21] Benedict, W. S., Gailar N., and Plyler E. K. *J. Chem. Phys.* 1956, 24, 1139–1165.
[22] Xantheas, S. *J. Chem. Phys.* 1996, 104, 8821–8824.
[23] Feller, D., and Dixon, D. A. *J. Chem. Phys.* 1996, 100, 2993–2997.
[24] Smith, D. E., and Dang, L. X. *J. Chem. Phys.* 1994, 100, 3757–3762.
[25] Spackman, M. A. *J. Chem. Phys.* 1986, 85, 6579–6585.
[26] Spackman, M. A. *J. Chem. Phys.* 1986, 85, 6587–6601.
[27] *Spravochnik khimika* (Chemist's Handbook); Nikol'skii, B. P.; Ed.; Khimiya: Leningrad, 1971; Vol. 1.
[28] Haile, J. M. *Molecular dynamics simulation. Elementary methods*; John Wiley & Sons: New York, 1992.
[29] Bartell, L. S. *J. Phys. Chem.* 1997, 101, 7573–7583.
[30] Koshlyakov, V. N. *Zadachi dinamiki tverdogo tela i prikladnoi teorii giroskopov* (Problems in Dynamics of Solid and Applied Theory of Gyroscopes); Nauka: Moscow, 1985.
[31] Sonnenschein, R. *J. Comp. Phys.* 1985, 59, 347–350.
[32] Landau, L. D., and Lifshitz, E. M. *Electrodinamica sploshnykh sred* (Electrodynamics of Continuous Media), Vol. 8.; Nauka: Moscow, 1982.
[33] *Fizicheskaya entsiklopediya* (Physical Encyclopedia), Vol. 1.; Prokhorov, A. M.; Ed.; Sovetskaya Entsiklopediya: Moscow, 1988.
[34] Bresme, F. *J. Chem. Phys.* 2001, 115, 7564–7574.
[35] Neumann, M. *J. Chem. Phys.* 1985, 82, 5663–5672.
[36] Stern, H. A., and Berne, B. J. *J. Chem. Phys.* 2001, 115, 7622–7628.
[37] Klein, R. A. *J. Comp. Chem.* 2002, 23, 585–599.
[38] Klein, R. A. *J. Amer. Chem. Soc.* 2002, 124, 13931–13937.
[39] Klein, R. A. *J. Comp. Chem.* 2003, 24, 1120–1131.

[40] Galashev, A. Y., Rakhmanova, O. R., and Chukanov, V. N. *Rus. J. Phys. Chem.* 2005, 79, 1455–1459.
[41] Galashev, A. Y., Chukanov, V. N., Novruzov, A. N., and Novruzova, O. A. *High Temp.* 2006, 44, 364–372.
[42] Novruzov, A. N., Chukanov, V. N., Rakhmanova O. R., and Galashev A. Y. *High Temp.* 2006, 44, 932–940.
[43] Galashev, A. Y., Rakhmanova, O. R., Galasheva, O. A., and Novruzov, A. N. *Phase Transitions.* 2006, 79, 911–920.
[44] Galasheva, A. A., Rakhmanova, O. R., Novruzov, A. N., and Galashev, A. Y. *Colloid. J.* 2007, 69, 56–65.
[45] Galashev, A. Y., Chukanov, V. N., Novruzov, A. N., and Novruzova, O. A. *Rus. J. Electrochem.* 2007, 43, 136–145.
[46] Chukanov, V. N., and Galashev, A. Y. *Dokl. Phys. Chem.* 2008, 421, Part 2, 226–229.
[47] Galashev, A. Y. *Environ. Chem. Lett.* 2009, DOI 10.1007/s10311-009-0243-9.
[48] Neumann, M. *J. Chem. Phys.* 1986, 85, 1567–1580.
[49] Angell, C. A., and Rodgers, V. *J. Chem. Phys.* 1984, 80, 6245–6252.
[50] Goggin, P. L., Carr, C. In Book *Water and Aqueous Solutions*; Neilson, G. W., and Enderby, J. E.; Ed.; Adam Hilger: Bristol–Boston, 1986; Vol. 37. pp 149–161.
[51] Hertzberg, G. *Raman and Infrared Spectra of polyatomic Molecules, Molecular Spectra and Molecular Structure*; Van Nostran: New Jersey, 1945.
[52] Kozintzev, V. I., Belov, M. L., Gorodnichev, V. A., Fedotov, Yu. V. *Lazernyi optiko-akusticheskii analis mnogokomponentnyh gazovyh smesei* (Laser Optical Accustic Analysis of Multicomponent Gas Mixtures); The MSTU named N.E. Bauman Press: Moscow, 2003.
[53] Poulet, H., and Mathieu, J.-P. *Specters de Vibration et Symetrie des Cristaux*; Gordon and Breach: New York, 1970.
[54] Daniel, J. S., Solomon, S., Sanders, R. W., Portmann, R. W., and Miller, D. C. *J. Geophys. Res.* 1999, 104, 16785–16791.
[55] Galashev, A. Y., and Rakhmanova O. R. *Colloid. J.* 2009, 71, 745–753.
[56] Galashev, A. Y., Rakhmanova, O. R., and Novruzova, O. R. *Rus. J. Gen. Chem.* 2009, 79, 1765–1772.
[57] Galashev, A. Y. *Rus. J. Phys. Chem. A.* 2009, 83, 2249–2254.
[58] Lemberg, H. L., and Stillinger, F. H. *J. Chem. Phys.* 1975, 62, 1677–1690.

[59] Rahman, A., Stillinger, F. H., and Lemberg, H. L. *J. Chem. Phys.* 1975, 63, 5223–5230.
[60] Saint-Martin, H., Hess, B., and Berendsen, H. J. C. *J. Chem. Phys.* 2004, 120, 11133–11143.
[61] Perera, L., and Berkowitz, M.L. *J. Chem. Phys.* 1991, 95, 1954–1963.
[62] Hunt, S. W., Roeselova, M., Wang, W., Wingen, L. M., Knipping, E. M., Tobias, D. J., Dabdub, D., and Finlayson-Pitts, B. J. *J. Phys. Chem. A* 2004, 108, 11559–11572.
[63] Berendsen, H. J. C., Postma, J. P. M., van Gunsteren, W. F., DiNola, A., and Haak, J. R. *J. Chem. Phys.* 1984, 81, 3684–3990
[64] Bosma, W. B., Fried, L. E., and Mukamel, S. *J. Chem. Phys.* 1993, 98, 4413–4421.
[65] Huiszoon, C. *Mol. Phys.* 1986, 58, 865–885.
[66] Potapova, G. F., Klochikhin, V. L., Putilov, A. V., Kasatkin, E. V., and Kozlova, N. V. In Book *Proc. of the 1st All-Russ. Conf. on Ozone and Other Ecological Pure Oxidants. Science and Technology*; Mosk. State. Univ.: Moscow, 2005.
[67] Murphy, W.F. *J. Chem. Phys.* 1977, 67, 5877–5882.
[68] Reid, P.J. *Acc. Chem. Res.* 2001, 34, 691–698.
[69] Brooksby, C., and Prezhdo, O. V., Reid, P. J. *J. Chem. Phys.* 2003, 118, 4563–4572.
[70] Upschulte, B. L., Green, B. D., Blumberg, W. A., and Lipson, S. J. *J. Phys. Chem.* 1994, 98, 2328-2336.
[71] Vallee, P., Lafait, J., Ghomi, M., Jouanne, M., and Morhange, J. F. *J. Molec. Struct.* 2003, 651–653, 371-379.
[72] Goldschleger, I. U., Kerenskaya, G., Janda, K. C., and Apkarian, V. A. *J. Phys. Chem. A* 2008, 112, 787-789.
[73] Andrews, L., and Spiker, R. C. Jr. *J. Phys. Chem.* 1972, 76, 3208-3213.
[74] Vostrikov, A. A., Dubov, D. Yu., and Drozdov, S. V. *Tech. Phys. Lett.* 2008, 34, 221–224.
[75] Danilov, A.D. and Karol', I.L. *Atmosfernyi ozon. Sensatsii i real'nost'*(Atmospheric Ozone. Sensations and Reality); Gidrometeoizdat: Leningrad, 1991.
[76] Schenter, G.K., Kathmann, S.M., and Garrett, B.C. *Statistical Mechanics of Water Clusters* (http://www.emsl.pnl.gov/docs/annual_reports/tms/annual_report1999(1619b-6t.html).
[77] Gonzales, E.Kh., Poltev, V.I., Teplukhin, A.V., and Malenkov, G.G. *Zh. Strukt. Khim.*, 1994, 35, 113–121.
[78] Mhin, B. J., Lee, S.J., and Kim, K.S. *Phys. Rev. A,* 1993, 48, 3764–3770.

[79] Galashev, A.Y., Rakhmanova, O.R., and Chukanov, V.N. *Khim. Fiz.*, 2005, 24, 90–96
[80] Galashev, A.Y., Rakhmanova, O.R., and Chukanov, V.N. *High Temp.* 2006, 47, 342–351.
[81] El'yashevich, M. A. *Atomnaya i molekulyarnaya spektroskopiya* (Atomic and Molecular Spectroscopy); Gos. Izd. Fiz._Mat. Lit.: Moscow, 1962.
[82] Novruzova O.A., Galasheva, A.A., and Galashev, A.Y. *Colloid Journal*, 2007, 69, 474–482.
[83] Novruzova O.A., Galasheva, A.A., and Galashev, A.Y. *Colloid Journal*, 2007, 69, 483–491.
[84] Novruzova O.A., and Galashev, A.Y. *High Temp.*, 2008, 46, 60–68.
[85] Zalogin, G. N., Kozlov, P. V., Kuznetsova, L. A., Losev, S. A., Makarov, V. N., Romanenko, Yu. V., and Surzhikov, S. T. In Book *Proceedings of the Third European Symposium on Aerothermodynamics for Space Vehicles* (*Noordwijk, The Netherlands, 24–26 November1998*); ESTEK: Noordwijk, 1998; pp 437–444.
[86] Galashev, A.Y. *Int. J. Sci. Eng.* 2009, 1, 31–38.
[87] Ghan, S. J., Leung, L. R., Easter, R. C., and Abdul-Razzak, H. *J. Geophys. Res.* 1997, 102, 777–794.
[88] Kebabian, P. L., Kolb, C. E., and Freedman, A. *J. Geophys. Res.* 2002, 107, 4670 (1–14).
[89] Hong, S. Y., Dudhia, J., and Chen, S. H. *Mon. Wea. Rev.*, 2004, 132, 103–120.
[90] Fletcher, N. H. *The Physics of Rain Clouds*; University Press: Cambridge, 1962.
[91] Betts, A. K., Ridgway, W. *J. Atmos. Sci.* 1988, 45, 522–536.
[92] Pierrehumbert, R. T., Brogniez, H., Roca, R. *The Global Circulation of the Atmosphere*; Schneider, T., and Sobel, A. H.; Eds.; Princeton University Press: Princeton, New York, 2006, pp.143–185.
[93] Gerding, M., Ritter, C., Neuber, R. *NPI Report Series*, 2002, 10, 77–80.
[94] Halmann, M. M., and Steinberg, M. *Greenhouse Gas Carbon Dioxide Mitigation. Science and Technology*; Lewis publishers: Roca Raton, London, New York, Washington, 1999; pp 7–8.

INDEX

A

absorption, viii, 2, 11, 12, 15, 17, 18, 20, 26, 28, 30, 32, 35, 36, 37, 41, 42, 44, 48, 50, 51, 52, 53, 60, 63, 64, 65
absorption spectra, 17, 28, 30, 36, 53, 64
adsorption, 35, 39
aerosols, viii, 4, 6, 44
aggregation, 4
algorithm, 27
aluminum oxide, 46
argon, 27, 53
atoms, 4, 9, 10, 23, 24, 25, 27, 35, 54

B

barium, 46
bending, 30
biosphere, 3
bromine, ix, 4, 23, 25, 27, 35, 37, 39, 64

C

carbon, viii, 1, 2, 3, 4, 9, 44, 46, 63
carbon dioxide, viii, 1, 3, 4, 44, 46, 63
catalyst, ix
catalytic effect, 6
cesium, 46
character, 49
chloral, 64
chlorine, ix, 4, 23, 25, 27, 30, 32, 46, 64
climate, vii, viii, 1, 59, 60, 63, 64, 65
climate change, viii, 59, 64
clustering, 64
clusters, viii, 3, 4, 5, 6, 9, 10, 11, 13, 14, 15, 17, 18, 19, 20, 23, 25, 26, 27, 39, 41, 42, 43, 45, 46, 47, 48, 51, 53, 55, 56, 58, 60, 61, 64, 65
combustion, 54
composition, 2, 5, 6, 46, 63
compressibility, 10
condensation, 5
conduction, 45
configuration, 10, 25
conservation, 18
consumption, 58
cooling, viii, 2, 3, 4, 5, 45, 64, 65
correlation, 14, 15, 47
cosmic rays, 2, 23
Coulomb interaction, 7, 25
covering, 52
crystals, viii, 18, 52, 56, 57, 58, 59, 60, 61, 65
cycles, 2, 64

D

danger, vii
decomposition, 28, 32

degradation, 47
deposits, viii, 5
desorption, 35
desorption of water, 35
destiny, vii
destruction, ix, 11, 64
detection, 5, 48
deviation, 49
dielectric permittivity, 15, 16
diffusion, 32, 43
dipole moments, 9, 28, 37, 39
discharges, 46
dispersion, 9, 13, 15
displacement, 32
dissociation, 32
disturbances, 63
dynamics, 10, 24, 25, 47, 68

E

Easter, 71
electric charge, 5, 9, 12, 25
electric field, 8, 13, 37
electromagnetic, 4, 5, 13, 15, 17, 18
electromagnetic field, 15
electrons, 23, 43
emission, viii, 19, 32, 33, 39, 40, 53
empirical potential, 10
engineering, 46
equality, 24
equilibrium, 2, 10, 25, 54
equipment, 5
evaporation, 45, 54, 59, 65
excitation, 32, 53
extinction, 11

F

flooding, 1
fluctuations, 39
formula, 11, 12, 13, 56
fractures, 64
fragments, viii
free energy, 47, 48
freedom, 54
frequencies, 15, 18, 19, 24, 28, 33, 35, 39
frequency dependence, 13, 19
frequency distribution, 19

G

gamma rays, 4
graph, 2
greenhouse gases, viii, 1, 3, 5, 45, 65

H

halogen, 23
heat capacity, 43
heat transfer, 45, 46
human activity, viii
hydrogen, viii, 4, 7, 9, 13, 30, 46, 64
hydrogen bonds, 30
hydrogen chloride, 46

I

impacts, 1
impurities, 44, 63
incidence, 13
inclusion, 53
insight, 7
integration, 10, 56
interdependence, viii
intermolecular interactions, 25
ionization, 5, 23, 64
ions, ix, 23, 25, 27, 28, 30, 32, 34, 35, 36, 37, 39, 43, 46, 64
IR spectra, 5, 15, 18, 30, 35
issues, viii

L

liberation, 23
lifetime, 43
linear molecules, 20
liquid phase, 23, 52

lithium, 46
localization, 9, 18, 37
lying, 1, 9

M

magnetic field, 2
magnetism, 1
majority, 4
matrix, 26
media, 50
melt, 1
melting, 43
meter, 45
mixing, 59
moisture, 15, 49, 50, 52, 56, 57, 60, 63, 65
moisture content, 50
molecular dynamics, 25
molecular oxygen, 35
monomers, viii, 50, 51, 52, 56, 60
Moscow, 67, 68, 69, 70, 71

N

Netherlands, 71
nitrogen, 2, 9, 46, 53, 59, 63, 64
nitrous oxide, 1
nonequilibrium, 53
nucleation, 5
nuclei, 57
nucleus, 12, 64

O

oceans, vii, 1, 45, 50, 59
optical properties, 5
orbit, 1, 3
oscillations, 17, 28
oxygen, ix, 2, 4, 9, 13, 23, 25, 26, 27, 28, 30, 31, 32, 34, 35, 36, 37, 39, 53, 63, 65
ozone, ix, 1, 3, 6, 11, 23, 25, 27, 28, 30, 31, 32, 34, 35, 37, 39, 43, 44, 47, 63, 64, 65
ozone-oxygen mixture, 29

P

permittivity, 11, 12, 13, 15
phonons, 17, 18
photolysis, 23, 31
photons, 17, 59
physical properties, 7
physicochemical properties, 10
polarizability, 8, 9, 26, 39
polarization, 8, 9, 13
pollution, 4
power plants, 3
precipitation, 5
probability, 47
properties, viii, 4, 7, 10
pure water, 15, 20

Q

quantum chemistry, 6

R

radiation, viii, 2, 3, 4, 5, 6, 12, 15, 17, 18, 19, 20, 23, 26, 28, 32, 35, 39, 41, 43, 44, 46, 47, 48, 50, 51, 52, 54, 55, 60, 63, 64, 65
radicals, 23
radius, 10, 25, 49
radon, 23
rainfall, 45
Raman spectra, 13, 30, 31, 32, 38, 39, 64
reactions, 23
reagents, 46
reality, 28
recommendations, viii
red shift, 37
redistribution, 51
reflection, 1, 13, 20, 64
refractive index, 13
relaxation, 31, 32
reliability, 64
replacement, 2
resolution, 33

respect, 13, 48, 51, 57, 58
rotations, 10
roughness, 21

S

saturation, 57
scaling, 28
scatter, 44
scattering, 11, 13, 41, 44, 54, 63, 64
sensing, 56
shape, 32, 37
simulation, 9, 10, 25, 32, 48, 68
smog, ix
sodium, 46
solid phase, 3
solvation, 31
space, 1, 44, 46, 47, 63, 64, 65
Spring, 4
stabilization, vii, 9, 13
stretching, 13, 30
stroke, 27
strontium, 46
sulfuric acid, 6
Sun, 1, 3, 4, 44, 49, 63
susceptibility, 49
symmetry, 14

T

temperature, vii, viii, 1, 2, 3, 6, 10, 25, 28, 43, 46, 47, 57, 58, 59, 60, 61, 65
temperature dependence, 57, 58, 61
territory, 49
thermal energy, 54
thermodynamic equilibrium, 48
transformations, 35
transmission, 63
transparency, 15, 63
transparent medium, 13
transport, 2

U

uniform, 11

V

vacuum, 13, 32
vapor, viii, 2, 3, 5, 7, 44, 47, 52, 56, 57, 58, 59, 60, 61, 65
variations, 2
vector, 8, 12
vehicles, 47, 65
velocity, 11, 24, 28
vibration, 13, 41

W

water clusters, viii, 5, 6, 7, 9, 10, 11, 13, 14, 19, 20, 23, 41, 46, 47, 48, 50, 52, 54, 55, 58, 60, 64
water evaporation, 3, 65
water vapor, viii, 1, 2, 3, 5, 13, 43, 47, 51, 57, 59, 64, 65, 67
wavelengths, 4, 44
windows, 63

X

X-ray, 4, 43